GLOBAL MONITORING: THE CHALLENGES OF ACCESS TO DATA

UCL
PRESS

GLOBAL MONITORING: THE CHALLENGES OF ACCESS TO DATA

Ray Harris, BA, PhD
and
Richard Browning, BSc, MSc
University College London

UCL
PRESS

First published in Great Britain 2005 by UCL Press,
an imprint of Cavendish Publishing Limited, The Glass House,
Wharton Street, London WC1X 9PX, United Kingdom
Telephone: + 44 (0)20 7278 8000   Facsimile: + 44 (0)20 7278 8080
Email: info@uclpress.com
Website: www.uclpress.com

Published in the United States by Cavendish Publishing
c/o International Specialized Book Services,
5824 NE Hassalo Street, Portland,
Oregon 97213-3644, USA

Published in Australia by Cavendish Publishing (Australia) Pty Ltd
45 Beach Street, Coogee, NSW 2034, Australia
Telephone: + 61 (2)9664 0909   Facsimile: + 61 (2)9664 5420

British Library Cataloguing in Publication Data
Harris, Ray, Dr
Global monitoring: the challenges of access to data
1 Environmental monitoring – Data processing
2 Computers – Access control
I Title   II Browning, Richard
363.7'063'028558

ISBN 1-84472-024-1
ISBN 978-1-844-72024-8

1 3 5 7 9 10 8 6 4 2

Typeset by Phoenix Photosetting, Chatham, Kent

Printed and bound in Great Britain

FOREWORD

The Treaty of the European Union enshrines sustainable development as a central objective of the EU, and confirms the principle that environmental protection must be integrated into all European policies and activities. Over the coming years, progress on economic growth, social development and environmental quality is therefore to be pursued with both a balanced approach and a long-term perspective.

However, the preparation and successful implementation of sustainable development policies depends, to a very large extent, on the quality of the information base available to advise the decision-making processes. The actual effects of policies are often the result of complex interactions developing over long periods of time and their assessment requires the mobilisation of best knowledge. Furthermore, progress towards real sustainable development requires the joint effort of all societal actors, which can only be achieved if all stakeholders are properly informed. Good governance and public participation at large also rest on the ready availability of an adequate information platform.

Thus, access to information is a prerequisite to implementing successful, sustainable development policies. This book, however, demonstrates that access to information is generally not straightforward, that it remains quite difficult in a number of cases and that urgent action is therefore needed to address the problem.

As part of the work programme of the Initial Period of the European Global Monitoring for Environment and Security (GMES) Action Plan (2002–2003), the authors have conducted a detailed analysis of the obstacles to accessing information. They gathered facts and views from a wide range of institutions and specialists on information production, including the Data Policy Working Group of the GMES Steering Committee. Of particular interest was the experience of the teams in charge of the GMES Thematic Projects, who had to produce information products useful to EU policies in a short period of time. These projects were confronted with the actual issues of access to data and information. Datasets were difficult to localise, their usage was sometimes prohibited or the cost of purchasing the data could not be afforded. Often, data was not properly documented and, in some instances, datasets even appeared to have been lost. Many of these problems were found to have their roots in data policies, with wide differences between countries apparent. The absence of a proper and explicit formulation of these policies was also evident in many cases. The causes of the problems also lie in technical gaps, or wider socio-economic and institutional issues.

This book conveys the clear message that data policies should be put at the service of better access to information, and suggests courses of action to be considered in preparing the establishment by 2008 of a European capacity for Global Monitoring for Environment and Security.

We wish to express our warm thanks to the authors for this very competent and timely contribution to the European GMES initiative.

Michel Cornaert and Alan Edwards
European Commission
Directorate General, Research, Environment and
Sustainable Development Programme

ACKNOWLEDGMENTS

We would like to thank the following people for their assistance in preparing this book: David Briggs, Michel Cornaert, Nina Costa, Alan Edwards, Oliver Greening, Colin Hicks, Paul Longley, Peter Ryder, Zofia Stott, Wolfgang Schneider, Graham Thomas, Pam Vass and Barry Wyatt. We would also like to thank the European Commission for research funding support, and all the people who responded to our requests for information during the course of our research.

The authors and publishers would like to thank the following for allowing images and diagrams to be used in the text: Telegeography Research Group – PriMetrica Inc; International Development Research Centre, Canada; US Census Bureau; European Environment Agency; Deutsches Zentrum für Luft-und Raumfahrt (DLR); British Atmospheric Data Centre; National Snow and Ice Data Centre, US; Land Processes Distributed Active Archive Centre, Sioux Falls, US; Statistics Denmark; and Centre for Ecology and Hydrology – Natural Environment Research Council UK.

LIST OF CONTRIBUTORS

David Briggs is Professor of Environmental and Health Sciences at Imperial College, London. He has extensive experience of the use of environmental data and information in process studies.

Richard Browning has a BSc from Queen Mary, University of London and an MSc in Remote Sensing from the University of London. He was a member of the project team on the DPAG (Data Policy Assessment for GMES) project that formed the basis for this book.

Nina Costa is a senior consultant with ESYS plc, Guildford. She has considerable experience of the European space industry, in particular in the fields of satellite navigation and Earth observation. Before joining ESYS in 2001, she worked at the European Commission's Joint Research Centre in Ispra, Italy.

Oliver Greening is a consultant with ESYS plc. He has over six years' experience developing information strategies and services for predominantly public sector and not-for-profit organisations. Oliver has an MSc in Remote Sensing from the University of London, a BSc in Geography, and is in the final stages of an MBA from the Open University Business School.

Ray Harris is Professor of Remote Sensing and Dean of the Faculty of Social and Historical Sciences at University College London. He has worked in both industry and academia on Earth observation topics since 1976, including work on data policy with the European Commission and the European Space Agency.

Peter Ryder is a consultant with Environmental Information Services, providing advice on the modalities, conduct, organisation and value of operational information services, based on the sciences of oceanography, hydrology and meteorology. He was formerly the Director of Operations at the UK Meteorological Office.

Zofia Stott has worked in the areas of space systems and policy for over 20 years. Her work has included mathematical modelling, market analyses and policy impact studies for space agencies, government departments, private industry and the European Commission. She is an independent consultant, working closely with ESYS plc on projects associated with GMES.

Pam Vass is an independent consultant in Earth observation, space technologies and the environmental sciences. The work associated with the socio-economic benefits chapter of this book was performed on behalf of ESYS plc. Her specialist skills include technical documentation preparation, proposal development, and project and quality management. She has scientific interests in biodiversity, ecology, integrated conservation and development, and natural hazards risk management.

Barry Wyatt was Director of the Monks Wood establishment of the Centre for Ecology and Hydrology (UK Natural Environment Research Council) from 1996 to 2002. During 2003–04 he co-ordinated the BICEPS (Building an Information Capacity for Environmental Protection and Security) project to assess the scientific and technical aspects of the planning of GMES.

CONTENTS

LIST OF FIGURES

LIST OF TABLES

LIST OF ABBREVIATIONS

AATSR	Advanced Along Track Scanning Radiometer
ABCR	Archive to Benefit Cost Ratio
ACMAD	African Centre of Meteorological Applications for Development
ACRES	Australian Centre for Remote Sensing
ADEN	European node for the ALOS mission
AFG	Administrative and Finance Group
ALOS	Advanced Land Observing Satellite
ANSI	American National Standards Institute
ANZLIC	Australia and New Zealand Land Information Council
ASDD	Australian Spatial Data Directory
ASDI	Australian Spatial Data Infrastructure
ASTER	Advanced Spaceborne Thermal Emission and Reflection Radiometer
ATSR	Along Track Scanning Radiometer
AVHRR	Advanced Very High Resolution Radiometer
BADC	British Atmospheric Data Centre
BEQUALM	Biological Effects Quality Assurance in Monitoring Programmes
BGS	British Geological Survey
BICEPS	Building an Information Capacity for Environmental Protection and Security
BIOPRESS	Linking Pan-European Land Cover Change to Pressures on Biodiversity
BKG	Bundesamt für Kartographie und Geodäsie
BLS	Bureau of Labour Statistics
BODC	British Oceanographic Data Centre
CAFE	Clean Air For Europe
CBA	Cost Benefit Analysis
CBD	Convention on Biodiversity
CEC	California Energy Commission

CEO	Centre for Earth Observation
CEOS	Committee on Earth Observation Satellites
CEPR	Centre Européen de Prévention des Risques
CFSP	Common Foreign and Security Policy
CGDI	Canadian Geospatial Data Infrastructure
CGMS	Co-ordination Group for Meteorological Satellites
CLC	CORINE Land Cover
CLRTAP	Convention on Long-Range Transboundary Air Pollution
CNES	Centre National d'Etudes Spatiales
CoE	Council of Europe
COFORD	Irish National Council for Forest Research and Development
CORBA	Common Object Request Broker Architecture
CORINE	Co-ordination of Information on the Environment
CRED	Centre for Research on the Epidemiology of Disasters
CSIRO	Commonwealth Scientific and Industrial Research Organisation
DAAC	Distributed Active Archive Center
DCT	Digital Cassette Tape
DEFRA	Department for Environment, Food and Rural Affairs
DFD	Deutsches Fernerkundungsdatenzentrum
DG	Directorate General
DGRC	Digital Government Research Centre
DISMAR	Data Integration System for Marine Pollution and Water Quality
DLR	German Aerospace Centre
DMSP	Defense Meteorological Satellite Program
DMU	Danish national environment research institute
DOSTAG	Data Operations Scientific and Technical Advisory Group
DPAG	Data Policy Assessment for GMES
DTD	Document Type Definition
DVB	Digital Video Broadcast
ECHO	European Commission Humanitarian Aid Office
ECMWF	European Centre for Medium-Range Weather Forecasts

ECOMET	Economic Interest Grouping of National Meteorological Services of the European Economic Area
EDC	Energy Data Collection programme
EDMED	European Directory of Marine Environmental Data
EEA	European Environment Agency
EECCA	Eastern Europe, Caucasus and Central Asia
EEZ	Exclusive Economic Zone
EFTA	European Free Trade Association
EIA	US Department of Energy's Information Administration
EIONET	Earth Information and Observation Network
EMSC	European-Mediterranean Seismological Centre
ENSO	El Niño/Southern Oscillation
Envisat	European environmental satellite
EODC	Earth Observation Data Centre
EOLES	Earth Observation Linking SMEs to face real time natural disaster management
EOPOLE	Earth Observation Data Policy and Europe
EOS	Earth Observing System
EPS	EUMETSAT Polar System
ERS	European Remote Sensing Satellite
ESA	European Space Agency
ESAC	Earth Science Advisory Committee
ESC	European Seismological Commission
ESDI	European Spatial Data Infrastructure
ESDP	European Security and Defence Policy
ESIS	European Shared Information Service
ESONET	European Seafloor Observatories Network
ESS	European Statistical System
ETCs	European Topic Centres
EU	European Union
EUFOREO	EU Forum on Earth Observation Use for Environment and Security
EUMARSIN	European Marine Sediment Information Network

EUMETCast	Data Distribution System of EUMETSAT
EUMETNET	European Meteorological Services Network
EUMETSAT	European Organisation for the Exploitation of Meteorological Satellites
EUROCORE	European database of marine sediment cores
EUROPA	European Union online
EUROSEISMIC	European database of geophysical data derived from marine seismic surveys
EUROSION	Coastal erosion in Europe
Eurostat	Statistical Office of the European Union
EUSC	European Union Satellite Centre
EU-SEASED	European Union Sea Sediments Database
EWSE	European Wide Service Exchange
ExIA	Extended Impact Assessment
FAO	UN Food and Agriculture Organization
FDS	Framework Definition and Support
FEMA	US Federal Emergency Management Agency
FGDC	US Federal Geographic Data Committee
FPAR	Fraction of Photosynthetically Active Radiation
FSB	Federal Security Service, Russia
FTP	File Transfer Protocol
Galileo	European navigation satellite
Gbps	Gigabytes per second
GCOS	Global Climate Observing System
GEIN	German Environmental Information Network
GEIN	Geospatial Emergency Information Network
GEIXS	Geological Electronic Information Exchange System
GEO	Group on Earth Observation
GeoTIFF	Geocoded Tagged-Image File Format
GIS	Geographic Information System
GISCO	Geographic Information Service of the European Commission
GMES	Global Monitoring for Environment and Security

GMES-GATO	Global Atmospheric Observations
GMS	Geostationary Meteorological Satellite
GNI	Gross National Income
GNP	Gross National Product
GNSS	Global Navigation Satellite System
GOCE	Gravity Field and Steady-State Ocean Circulation Mission
GODAE	Global Ocean Data Assimilation Experiment
GOES	Geostationary Operational Environmental Satellite
GOME	Global Ozone Monitoring Experiment
GOOS	Global Ocean Observing System
GPCP	Global Precipitation Climatology Project
GRACE	Gravity Recovery and Climate Experiment
GRID	Global and Regional Integrated Data centres
GTOS	Global Terrestrial Observing System
GTS	Global Telecommunications System
HDF	Hierarchical Data Format
HDT	High Density Tape
HRI	High Resolution Imagery
IACG	Canadian Interagency Committee for Geomatics
IACMST	Inter-Agency Committee on Marine Science and Technology
ICP Forests	International Co-operative Programme on Assessment and Monitoring of Air Pollution Effects on Forests
ICSU	International Council for Science
IDEC	Spanish spatial data infrastructure organisation
IFRC	International Federation of the Red Cross
IFRC&RCS	International Federation of Red Cross and Red Crescent Societies
IGBP	International Geosphere Biosphere Programme
IGN	Instituto Geografico Nacional, Spain
IGN	Institut Geographique National, France
IGOS	Integrated Global Observing System
IMF	International Monetary Fund

INASP	International Network for the Availability of Scientific Publications
INCITS	International Committee for Information Technology Standards
INFEO	Information on Earth Observation
InSAR	Interferometric Synthetic Aperture Radar
INSEE	Institut National de la Statistique et des Études Économiques, France
INSPIRE	Infrastructure for Spatial Information in Europe
IOC	Intergovernmental Oceanographic Commission
IPCC	Intergovernmental Panel on Climate Change
IPR	Intellectual Property Rights
ISC	International Seismological Centre
ISO	International Standards Organization
ISOOS	Integrated Sustained Ocean Observing System
JERS	Japanese Earth Resources Satellite
JPEG	Joint Photographic Experts Group
JRC	Joint Research Centre
KNMI	Koninklijk Nederland Meteorologisch Instituut, the Netherlands
LADAMER	Land Degradation Assessment in Mediterranean Europe
LAI	Leaf Area Index
Landsat	US Earth resources satellite
LCC	Land Cover Change
LEDC	Less Economically Developed Country
LOIS	Land Ocean Interaction Study
MANHUMA	Utilization of Earth Observation for the Management and Conservation of Wetlands
MARS	Meteorological Archival and Retrieval System
Mbps	Megabytes per second
MERIS	Medium Resolution Imaging Spectrometer
MERSEA	Marine Environment and Security in the European Area
Meth-MonitEUr	Methane Monitoring for the European Region
MODIS	Moderate Resolution Imaging Spectroradiometer

MSG	Meteosat Second Generation
MSS	Multispectral Scanner
MTP	Meteosat Transition Programme
NARA	US National Archives and Records Administration
NASA	US National Aeronautics and Space Administration
NASDA	National Space Development Agency of Japan
NDSC	Network for the Detection of Stratospheric Change
NERC	Natural Environment Research Council
NetCDF	Network Common Data Format
NGO	Non-governmental organisation
NIMSA	National Interest Mapping Services Agreement
NLS	National Land Survey
NMCAs	European National Mapping and Cadastral Agencies
NMS	National Meteorological Service
NOAA	US National Oceanic and Atmospheric Administration
NPP	Net Primary Productivity
NRC	Natural Resources Canada
NRCS	US Natural Resources Conservation Service
NRT	Near-Real Time
NSDI	National Spatial Data Infrastructure
NSIDC	National Snow and Ice Data Center
NSLRSDA	US National Satellite Land Remote Sensing Data Archive
OCEANIDES	Harmonised Monitoring, Reporting and Assessment of Illegal Marine Oil Discharges
OCHA	UN Office for the Co-ordination of Humanitarian Affairs
ODISSEO	Open Distributed Information Systems and Services for Earth Observation
OECD	Organisation for Economic Co-operation and Development
OGC	Open GIS Consortium, renamed the Open Geospatial Consortium
ORFEUS	Observatories and Research Facilities for EUropean Seismology
OSCAR	Ocean Surface Current Analyses – Real time

OSCE	Organization for Security and Co-operation in Europe
OSPAR	Convention on the Protection of the Marine Environment of the North-East Atlantic
PAC	Policy Advisory Committee
PANGAEA	Network for geological and environmental data
PB-EO	Programme Board for Earth Observation
PDF	Portable Document Format
POP	Post Office Protocol
PSMA	Public Sector Mapping Agencies
PUMA	Preparation for the Use of MSG in Africa
QUASIMEME	Quality Assurance Laboratory Performance Studies for Environmental Measurements in Marine Samples
QuikSCAT	US Quick Scatterometer
Ramsar	International wetlands convention
RCS	Red Crescent Society
RESURS	Earth Remote Sensing System
RFA	Regional Forest Agreement
ROD	Reporting Obligations Database
ROI	Return on investment
RTD	Research and Technology Development
SABE	Seamless Administrative Boundaries of Europe
SAC	Special Areas of Conservation
SAF	Satellite Application Facility
SAR	Synthetic Aperture Radar
SCI	Sites of Community Importance
SDDS	Special Data Dissemination Standard
SDTS	Spatial Data Transfer Standard
SeaWiFS	Sea-viewing Wide Field of view Sensor
SIBERIA	Multi-sensor concepts for greenhouse gas accounting of northern Eurasia
SMEs	Small and Medium sized Enterprise
SNS	Danish Forest and Nature Agency
SPA	Special Protection Area

SPOT	Satellite pour l'Observation de la Terre
SRTM	Shuttle Radar Topography Mission
SSM/I	Special Sensor Microwave/Imager
STAR	Sustainability Targets And Reference database
STAR	Solar and Thermal Atmospheric Radiation project
STG	Scientific and Technical Group
Tacis	EU assistance to eastern Europe and central Asia
TESEO	Treaty Enforcement Services using Earth Observation
TM	Thematic Mapper
TMACS	TM/MSS Archive Conversion System
UNECE	United Nations Economic Commission for Europe
UNEP	United Nations Environment Programme
UNHCR	United Nations High Commission for Refugees
UNSD	United Nations Statistics Division
USGS	United States Geological Survey
VSAT	Very Small Aperture Terminal
WCRP	World Climate Research Programme
WDC	World Data Centre
WGP	Working Group on Data Policy
WGS84	World Geodetic System 1984
WHO	World Health Organization
WMO	World Meteorological Organization
WTO	World Trade Organization
XML	Extensible Markup Language

CHAPTER 1

INTRODUCTION AND OBJECTIVES

OBSERVATION AND MONITORING

For thousands of years, men and women have observed, measured and monitored their physical environment. The ancient Egyptians monitored the water levels in the River Nile because of their importance to agriculture; the ancient Greeks also had interests in monitoring the atmosphere, as shown by the Tower of the Winds in Athens; and Arab scientists, such as Ibn Khaldun and Ibn Battuta, contributed much to environmental observation (Izzi Dien 2000), including developing instruments to measure the environment. In many countries, regular meteorological observations started in the 19th century, at about the same time that serious mapping and measurement of the terrain began. Both meteorological monitoring and topographic mapping were stimulated by military requirements, often a stimulus for environmental information collection.

A crucial shift in the attempt to provide regular environmental measurement happened in 1960. On 1 April of that year, the first weather satellite was launched. While the early images were neither comprehensive nor of high quality, these satellites were the first steps along the way to providing synoptic, comprehensive environmental observations of the whole planet. One primary characteristic of the satellite remote sensing, or Earth observation, era is the desire and the ability to collect data for any part of the planet, desert or tropical rainforest, ice field or open ocean, Antarctica or France. That is not to say that all Earth observation satellites collect data over the whole globe, but rather the technical capacity of polar and geostationary orbits, combined with distributed downlinks, enables global coverage. However, Earth observation data is not enough by itself: it is also necessary to measure the environment *in situ* more fully. While total global data collection is not possible for many environmental parameters, there is a strong operational and scientific desire to improve ground-based measurement of, for example, atmospheric ozone, carbon dioxide, the net primary productivity of vegetation and ice mass balance.

The last few decades of the 20th century were characterised by a growing realisation of the impact of human beings on the global environment. Since the industrial revolution in Europe of the 18th and 19th centuries, the concentrations of greenhouse gases in the atmosphere have been increasing, largely as a result of burning fossil fuels. The effects of these atmospheric changes have now become sufficiently great for scientists and governments to express clear concerns for the future of our planet (IPCC 2001). So, at the start of the 21st century, we have both a need to monitor the environment globally and the technological capacity to do so. However, the challenges of accessing data and information on the global environment are many and are great. Our technological ability to monitor the environment has not yet been translated fully into an ability to access and use the data and information created by the technology.

INTERNATIONAL INITIATIVES

The issue of access to environmental and scientific data of our planet has a high scientific, technological and political profile. At the European level, the Sixth Environmental Action Programme, the Directive on Public Sector Information and the Aarhus Convention all illustrate the convergence of many policies on improving access to environmental data. The Göteborg Summit on Sustainable Development held in 2001, the World Summit on Sustainable Development held in Johannesburg in 2002 and recent meetings of the G8 ministers have all noted the need for the international community to monitor the environment, to improve our knowledge and understanding of environmental processes and to be able to predict future changes.

The concern over access to environmental data does not just involve Earth observation data, but a wide variety of other environmental and related data (Arzberger *et al* 2004). The International Council for Science (ICSU) has identified access to data as an important issue for all aspects of science and technology, while the European Commission Sustainable Development Strategy and the Sixth Environmental Action Programme both acknowledge the importance of data availability. The governments of 35 countries, meeting at ministerial level in the OECD Committee for Scientific and Technological Policy, declared their support for improvements in the access regimes for digital research data funded from public sources (OECD 2004).

Heads of state and prime ministers meeting at the G8 summit held in Evian, France, in 2003 (G8 2003) agreed an action plan on science and technology for sustainable development, with a focus on three areas that present great opportunities for progress: co-ordination of global observation strategies; cleaner, sustainable and more efficient energy use; and agricultural sustainability, productivity and biodiversity conservation. The strengthening of co-ordination on global observation strategies by the G8 countries has four main characteristics:

(1) To develop close co-ordination of the respective global observation strategies for the next 10 years and identify new observations to minimise data gaps.

(2) To build on existing work to produce reliable data products on atmosphere, land, fresh water, oceans and ecosystems.

(3) To improve the worldwide reporting and archiving of these data and fill observational gaps of coverage in existing systems.

(4) To favour interoperability with reciprocal data-sharing.

The discussion of improved co-ordination of global observing strategies was followed through in July 2003 at the Earth Observation Summit held in

Washington DC (EO Summit 2003). Governments from around the world committed themselves to greater co-ordination in Earth observation, and particularly on improving access to data. Europe played an important role in the Earth Observation Summit and has the opportunity to play a central role in future activities.

In the list of four issues identified by the G8 above, access to data is either implicit or explicit in all of them. Access to data involves both technical access and the policies that govern access, and therefore a concern for data policy is an essential part of the development of environmental monitoring. In Europe, this concern for global observation is finding one form of expression in the Global Monitoring for Environment and Security initiative.

GLOBAL MONITORING FOR ENVIRONMENT AND SECURITY (GMES)

First phase

The initiative on Global Monitoring for Environment and Security (GMES) is aimed at the establishment, by 2008, of a European capacity for the provision and use of operational information for the monitoring and management of the environment, and for civil security (EC 2001b). The first phase of GMES has contributed substantially to this long-term aim by understanding what is needed, by consolidating user requirements and by implementing an evolutionary approach to the development of GMES. The GMES EC Action Plan (EC 2001a), contained in a European Commission Communication published in 2001, recognised three key elements of work in the first phase of GMES:

(1) Delivering a set of pilot information products and services on priority environment and security topics.

(2) Assessing the current capabilities, organisational elements and policies required for the supply of information, and to determine how effectively they meet the user requirements.

(3) Elaborating the definition of the future system architecture and integrated services on the basis of user requirements.

The terms *global* and *security* in GMES have rather specific meanings. Global refers to the whole globe when regional development and humanitarian aid are considered because this is where the European Commission has what it calls competences, or agreed rights to act. For other sectors, such as climate change or marine pollution, the term *global* in GMES refers to the European contribution, along with other nations, to producing a global coverage of observations of environmental phenomena. However, when building space-based systems or services, there is typically an engineering requirement produced by the near polar orbit to collect information for the whole globe and not just to concentrate on information collection for Europe. The term *security* in GMES refers to civil security and includes topics such as drought and famine predictions, warnings of major earthquake activity, flooding and pollution events. If the European Union adopts a defence security element in the future at the political level, then the term *security* in GMES could be enlarged to encompass defence security. It is certain that much civil environmental information (including Earth observation) has defence applications, so GMES already has implicit links to dual use (civil and defence) applications.

The data offer in GMES

One way of summarising GMES that has been widely used is shown in Figure 1.1. It represents a way of integrating the wide variety of environmental and supporting data available, using the data in both models and also empirically to produce information that is required by the users in the information demand level of Figure 1.1.

An important part of GMES lies in ensuring that the wide variety of information summarised in the data provider level can be exploited in a coherent fashion. Different approaches to legal protection, data formats, standards, metadata, distribution, pricing and archiving mean that there are often obstacles to the successful use of a variety of Earth observation data and other environmental information in application projects. It is essential for the use of the data and the production of information that the data in GMES can be used coherently to achieve effective environmental monitoring.

OBJECTIVES OF THE BOOK

In the meteorology sector, global monitoring of the environment has been a reality for decades. Global monitoring of the land and the oceans is now an objective of

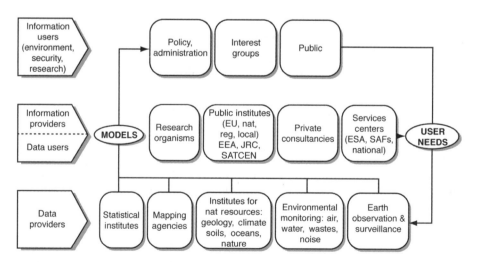

Figure 1.1: The overall structure in GMES for fostering the dialogue between the data suppliers and the information users

several organisations, for example the European Space Agency (ESA), the US National Aeronautics and Space Administration (NASA) and the US National Oceanic and Atmospheric Administration (NOAA). While satellite Earth observation data is highly valuable for global environmental monitoring, it is not the only data. What characterises new thinking about monitoring our environment is the need to incorporate a wide variety of datasets into operational monitoring programmes. If it were possible to perform environmental monitoring entirely from space, then the problems would be minimised: it is because of the wide diversity of data sources that it is vital to examine the difficulties that arise when using environmental data with different data policies. Even if data were only employed from NASA and ESA, there would be data policy conflicts: when data from statistical institutes, mapping agencies and environmental monitoring organisations is used together with Earth observation data, then the data policy difficulties and obstacles multiply.

This book arises out of work carried out in the context of the European initiative on Global Monitoring for Environment and Security, but the issues discussed are not restricted to GMES itself. The book tackles the questions of access to data for global environmental monitoring in the round, and to some extent the terms *GMES* and *global environmental monitoring* are used interchangeably. While GMES has a European focus, this book takes a global approach. The issues raised are certainly valid for Europe, but they are equally valid to the development of operational environmental monitoring in many nations including the US, Canada, South America, India, China, Japan and Australia. Likewise, the term *data policy* is a useful shorthand for the long list of issues about access to environmental data and information.

This book grew out of a project entitled Data Policy Assessment for GMES (DPAG). The project was funded by the European Commission Framework V research and development programme and was an integral part of the initial phase of GMES. At the same time as the DPAG project was in operation there were other parallel and related project activities in GMES that were cross-cutting in nature, that is, they looked across the application themes of GMES projects. The parallel activities were concerned with identifying gaps in knowledge, technology and tools, the adequacy of monitoring, and socio-economic and institutional issues. Some of the material from this parallel work has been included in this book as separate chapters (Chapters 8, 10 and 11).

The overall aim of this book is to contribute to the broad challenges of the development of global environmental monitoring systems for planet Earth, and specifically to contribute to the European programme of GMES. *Inter alia*, the book also aims to contribute to the wider international debate on environmental and scientific data management (Arzberger *et al* 2004). The objectives of this book are as follows:

(1) To identify and document the data policy issues and obstacles found when using information from environmental data providers. This corresponds to the data providers line in Figure 1.1.

(2) To identify the practical problems that arise from uncertainty or obscurity in environmental data policies. This is achieved by reference to GMES thematic projects and other application project experience in various parts of the world.

(3) To assess the impacts of data policies on access to data, information and information services.

(4) To analyse the limitations and the opportunities provided by the frameworks in which global environmental monitoring is evolving.

(5) To recommend ways forward to improve environmental information integration, with a strong emphasis on the user perspective.

Data policy should be the servant of the mission objectives: seen from a data supplier perspective, this is relatively straightforward; however, seen from a user perspective, where the user needs access to data from a large number of sources, then the servant becomes difficult and sometimes impossible to manage. So far, many environmental monitoring systems have been supplier-led: one key challenge for the new initiatives in global environmental monitoring is to allow the systems to be user-led. This will take bravery on the part of the data supplier organisations.

STRUCTURE OF THE BOOK

Global monitoring of the environment covers a wide variety of spatial scales. The concerns of accessing environmental data also cover a variety of scales and of public policies. This book contributes to the development of improved global monitoring by paying attention to the challenges of accessing these different scales of environmental data and information. After this introductory chapter, the book examines the following issues.

Chapter 2 provides a platform by documenting the data policies of the five main categories of organisation examined in this book, namely statistical institutes, mapping agencies, institutes for natural resources, environmental monitoring organisations and Earth observation organisations. The work in Chapter 2 is extended in Chapter 3 by an examination of the particular case of socio-economic data.

Having established the platform of data policies, Chapter 4 provides evidence for the practical challenges that real users face in trying to access environmental data, working particularly with evidence from projects sponsored by the European Commission and the European Space Agency. These experiences are supported by Chapter 5, which gives actual examples of problems of accessing environmental data for oil spill monitoring, earthquake monitoring and climate change analysis.

Chapter 6 reviews good practice on data policy for environmental data, drawing on evidence from Australia, New Zealand, Canada and the US.

Chapter 7 opens an examination of structural and technological issues by reviewing access to the Internet and the role of portals in global environmental monitoring. Chapter 8 follows on by examining the structural challenges of interoperability, linkage and data access: the chapter is written by David Briggs, Peter Ryder and Barry Wyatt. Chapter 9 looks at the access issues for a possible European Shared Information Service that could be constructed to provide European users with global environmental data.

Chapter 10, written by Nina Costa, Oliver Greening and Zofia Stott, widens the scope of the discussion to include the institutional factors that help to structure access to data. In Chapter 11, Pam Vass examines the socio-economic benefits that come from improved global monitoring.

Chapter 12 concludes the book, discusses sustainable funding and provides recommendations on future developments to improve data access for global environmental monitoring.

INFORMATION

The information used in this book has largely been collected from organisations in the data provider line of Figure 1.1, namely:

- statistical institutes;

- mapping agencies;

- institutes for natural resources: geology, climate, soils, oceans;

- environmental monitoring and regulation organisations; and

- Earth observation organisations.

During 2003 we contacted 167 organisations in these five categories to request information on their data policies. Of these 167, replies were received from 67 organisations. In addition, data policy information was gathered from websites or additional sources for other key organisations, making a total of 81 organisations, or about half of the targeted contacts. The organisations that provided information in response to our request, or for which we gathered information from other sources, are listed in Appendix A. The work reported in this book also draws on information collected by:

- direct contact and discussions with the main organisations in Earth observation and environmental monitoring in Europe;

- discussions with 14 GMES environmental applications projects run by the ESA and the European Commission;

- discussions and debates on GMES held at major European forums held in Brussels, Noordwijk, Athens and Baveno during 2002 and 2003;

- meetings with GMES staff in ESA and the European Commission, plus members of working groups drawn from European Member States working with the core GMES staff; and

- meetings with experts on spatial data infrastructure in Australia, New Zealand and Canada.

All these sources provided information, opinion and viewpoints on data policy and the challenges of accessing environmental data and information. But what is meant by data policy? We have adopted the following six characteristics of data policy as a template to assist in collecting and presenting information on data policy, and in developing ways to overcome obstacles in the data policies of different environmental monitoring organisations:

(1) ownership, privacy and confidentiality;

(2) intellectual property rights and associated legal frameworks;

(3) standards and metadata;

(4) licensing, distribution and dissemination;

(5) pricing policy; and

(6) archiving policy.

These six characteristics seem to cover the main concerns of policy or access in the environmental data domain.

CONTEXT

Introduction

Access to data and the availability of data affects many different fields and disciplines. The importance of data both for research and development and for operational applications is clear, and the issue of access to it is a subject that has received much attention in recent years. Some authors argue that scientific data should be archived and made available to all under international regulation (Dittert *et al* 2001); others are concerned that freedom of information legislation, such as that in the US, may be costly to university scientists and question their own independence (Macilwain 2000). The literature covering such issues is discussed in this section using four main categories: government information, mapping agencies, environmental monitoring organisations and institutes for natural resources.

Government information

Many governments have become increasingly keen to distribute widely the data and information they hold. Muir and Oppenheim (2002) argue that access to government-held information is essential to secure the proper level of knowledge in the general population about the democratic process, and Tate (1997) adds that access to government information is not only a right but also a necessity for the peaceful working of a nation or institution. Freedom of information legislation is being increasingly adopted by many countries to facilitate access to government information and statistics, whilst ensuring that individuals are aware of, and have control over, personal data.

In general, there is a steady trend towards governments disseminating more and more information, and reducing the barriers impeding access to the information that is held but not disseminated (Muir and Oppenheim 2002). However, the type of information and the freedom with which it is released varies between different countries. For example, Canada includes correspondence, memoranda, books, plans, photos, films and sound recordings under its freedom of information act, granting access to information regardless of its physical form and characteristics. Muir and Oppenheim note that freedom of information in the European Union varies between Member States, and in general the European Commission has moved towards freedom of information with a certain amount of reluctance. This has led to several legal actions being taken against the European Commission for its unwillingness to release information. It has also been noted that moves towards increasing freedom of information have slowed since 11 September 2001 because of concerns about terrorism.

Freedom of information and the ability of the public to gain statistical information from the government is clearly important, but such policies raise concerns among many for the privacy of individuals. Tate (1997) notes that individual privacy was the single most important issue according to Canadians, above topics such as national unity, unemployment or social benefits. Therefore, at the same time as the Canadian Access to Information Act was passed, the Privacy Act was also passed by the Canadian government. Such concerns have been clearly addressed by many national statistical organisations, ensuring that data is freely available but contains no information on individuals or individual businesses.

Whilst many believe that scientific data relating to the environment should be available on a free and unrestricted basis, there are also others who believe that such a policy can be detrimental to scientific research. For example, in the US, scientists have opposed the Freedom of Information Act that gives access to university researchers' data, arguing that it would lead to the harassment of researchers in sensitive fields. An example cited in the journal *Nature* (Macilwain 2000) described how the US Chamber of Commerce took the Environmental Protection Agency to court following three requests for the data from university studies on the sulphur content of petrol, the health effects of particulate matter and the vulnerability of poor people to pollution. University scientists are concerned that requests arising from the Freedom of Information Act will be costly and may put their independence at risk. The US Chamber of Commerce argues against this view, stating that publicly-funded research work should be available to the public.

As the number of households with access to the Internet increases, statistical information is frequently made available via the world wide web. However, others have noted that it can be difficult to make productive use of statistical data available on the web, and making this information accessible and useful has posed two major challenges (Ambite *et al* 2001). The first challenge is the integration of large, dispersed collections of data compiled by different people at different times and for different purposes. The second is overcoming the limitations of the web's browser paradigm to disseminate complex information derived from multiple sites.

Mapping agencies

Maps and spatial data provide important information about the environment, and consequently the accessibility of such data can have significant implications for society. The importance of accurate, freely available mapping and cadastral information is highlighted in a case study by Garciá and Tuladhar (2001) in Guatemala. Land issues are key to poverty alleviation in Guatemala. The issues include land property rights, which are unclear for most of the rural population, and land distribution. The registration of land titles and mortgages has been the responsibility of the Guatemalan Land Registry since 1877. However, these titles have no legal frame of reference and their quality has therefore gradually become less secure. There are also no regulations defining by law which national institution is responsible for cadastral activities, and no official cadastral registration exists in Guatemala. The lack of such data means that issues such as boundary disputes are difficult to resolve, with individual plans not referenced to any common system of designated geographical points. Efforts are now being made towards the establishment of a networked, distributed cadastral infrastructure, with the emphasis on inter-organisational partnerships to maximise data sharing and minimise data duplication. This data and information can then be acquired through the network of the cadastral infrastructure by users such as municipalities, land funds and those involved in conflict resolution.

Many have argued that the cost of map data can prohibit its use, to the extent that it has a negative impact upon wider society. For example, Klinkenberg (2003) argues that the true cost of spatial data is more than the data acquisition cost, but a combination of direct dollar costs and indirect social costs relating to access to and use of map data. Klinkenberg examines the Canadian case, where the Crown maintains copyright on all federally and provincially produced spatial data, and collects royalties on the data it sells to resellers, a situation similar to that in the UK. By comparison, no such copyright on data produced by the federal government exists in the US, and data are available for the cost of reproduction only. The US government regards its role as the provider of free and easy access to digital spatial data so that society can benefit from the increases in efficiency and equity that such access permits. Klinkenberg argues that the US government appears to take the view that there is a much greater economic benefit to be gained by making spatial data freely available than by charging for spatial data. Others suggest that state or Crown copyright by comparison 'has simply become a mechanism to create an artificial rarity value for mapping data, with the aim of covering the agency's costs' (Barr 1998). The New Zealand government maintained a similar policy until recently, retaining strict control over the distribution of spatial data through the imposition of Crown copyright and by charging high royalties. The New Zealand government has since changed its position, and royalty-free data is now available for the marginal cost of distribution.

It is also argued that the restrictive copyright policy followed by countries such as Canada and the UK has led to a distinct class system (Klinkenberg 2003), where data is only available to the select few who can afford to pay the price. This can have several negative impacts upon society. For example, Canadian academic research in areas such as wildlife habitat modelling, ecosystem analysis and social inequalities is affected because of the restrictive and often overwhelming cost of obtaining high-resolution, digital spatial data. Similarly, Canadian students using Geographic Information Systems (GISs) quickly discover that application examples are often derived from US jurisdictions, since US data is more freely available.

Standards for geospatial information and map data is another topic that is often discussed in the literature, particularly with regard to international standards and metadata. Houlding (2001) notes that there are a large number of proposed standards for GIS data, which essentially means that there is no real standard in the strict meaning of the term. Many of the existing standardisation efforts are dominated by efforts to prolong the life legacy of computer systems rather than to ensure the interoperability of GIS systems. In addition, despite the volume of data processed, there have been no concerted efforts to develop comprehensive metadata standards in the geosciences. There are two exceptions to this: the Federal Geographic Data Committee (FGDC) in the US and the Australia and New Zealand Land Information Council (ANZLIC), both of which have developed similar content standards for geospatial metadata. However, Houlding argues that both these standards have limitations. The FGDC standard is limited to GIS datasets and appears overly complex, employing a total of 334 different data descriptors. By comparison, the ANZLIC standard is primarily concerned with information regarding dataset accessibility and quality rather than the dataset itself. Houlding argues that both are, by design, closer in function to information retrieval systems as opposed to operational data standards.

Environmental monitoring organisations

Complex environmental problems such as global warming require data from a wide variety of environmental disciplines, and the magnitude and consequent volume and diversity of data needed highlights a number of major issues relating to the use and management of data.

An example project that illustrates the challenges is the UK Natural Environment Research Council's (NERC) Land Ocean Interaction Study (LOIS), which was concluded in 1998. Roberts and Moore (1998) looked at the main data policy issues that affect the LOIS project, which include:

- the volume and variety of datasets;
- harmonisation of similar data across different disciplines of science or across organisations;
- provision of access to data;
- balancing the rights of the collector and the funding agency;

- observing copyright and intellectual property rights;

- providing online metadata catalogues;

- providing direct remote access and removing the need for staff to service requests from remote users; and

- improving the user interface.

One of the key aims of LOIS was to leave policy makers and planners in a position to develop operational decision support systems to help them achieve sustainable management practices. The data management policy for LOIS, which is defined by the overall NERC policy, would therefore be key to achieving this aim. Roberts and Moore note that this policy created a number of considerable practical challenges for LOIS data in the form of: the diversity of data; the volume of data; a lack of standards; the number of suppliers; ensuring that intellectual property rights and copyright arc upheld; maintaining quality assurance and audit trails; maintaining security and confidentiality; finite resources; and the demand for data from a geographically dispersed user base with widely varying computer skills.

These challenges were met by developing organisational structures and new systems. With respect to the data, this was achieved through the establishment of five LOIS data centres, the main objectives of which were to identify user needs, acquire the data from various sources and disseminate the data to users. In addition, the data centres have the added responsibility of setting format standards, ensuring long-term security and managing the budget for the acquisition and dissemination of the identified datasets. Users can gain access to the data through the Internet, with the LOIS data system creating and maintaining its own metadata catalogue in the form of a series of web pages (Roberts and Moore 1998).

Other similar projects to aid adequate information handling and data management for global environmental change research have also been developed in recent years, and like LOIS these have also raised a number of issues about data access and data policy. A good example of such a project is PANGAEA, a network for geological and environmental data initiated in 1993. PANGAEA is an information system for the long-term archiving and publication of any georeferenced data related to the environmental sciences, and was conceived to be not just a data archive but a scientific tool (Diepenbrook *et al* 2002). The project was funded by the German Ministry of Research and Technology, initially to provide data management services on a national level in Germany, but was later expanded to include European Commission and international projects. A project on such a scale as PANGAEA raises a number of important data policy points, particularly regarding standards and metadata. Consequently, the data managers who are responsible for each dataset must ensure that the agreed standards are met when data is imported into the system. During the data-import process in PANGAEA, metadata is checked for completeness and validity. It is not only essential to have excellent quality datasets, it is vital that the user can estimate the quality of the data. The completeness of this metadata, including, in particular, descriptions of analytical methods and references to where data has first been published, is crucial in understanding the analytical data held by PANGAEA.

Similar data access and policy issues are raised by the Global Ocean Observing System (GOOS), another example of an ongoing, multi-disciplinary observing system focused on the production and delivery of ocean data and products to a wide variety of users. Nowlin *et al* (2001) have argued that for any sustained, integrated observing system such as GOOS the dataset must display certain characteristics. First, the dataset must be long-term and, once begun, measurements should continue into the foreseeable future. Continuity, they argue, is sought in the observed quantity, not necessarily in the measurement method. Secondly, data should be systematic and relevant and meet agreed standards: measurements should be made in a rational fashion, with defined spatial and temporal sampling, precision and accuracy characteristics combined with care in calibration.

Dittert *et al* (2001) note that the necessary database infrastructure to allow access to raw environmental data was created as early as the 1950s, when the International Council of Scientific Unions (ICSU, now named the International Council for Science) set up the World Data Centre (WDC). However, there are no international regulations requiring scientists to store results as raw data and accompanying metadata in this or any other publicly accessible archive. Dittert *et al* argue that an internationally binding regulation that adheres to WDC principles is required, and should guarantee that all scientific data is archived and freely available.

Institutes for natural resources

One example of a natural resources dataset was developed by Australia's Regional Forest Agreement (RFA) programme, a national initiative to plan forest use in key areas over 20 years (Bugg *et al* 2002). The project's aims were to develop a first class conservation reserve system for Australia's forests to maintain regional, environmental and social values, lay the foundations for ecologically sustainable management of production forests, and to secure access to timber resources for sustainable, internationally competitive industries over the ensuing 20 years. The scale of the project highlighted several data policy and data management issues, for example the requirement to ensure that data met the appropriate standards and was readily accessible for various purposes at multiple locations. Many RFA datasets were populated with historical data collected by different state agencies for different purposes, and data therefore met a range of original specifications. This meant that the data often had to be converted to consistent formats and standards for use with other data and modelling tools. Bugg *et al* note some of the main issues regarding data standards, including:

- size, complexity and number of datasets;
- different scales, resolution and reference systems;
- different attributes in different regions, for example different definitions for land tenure and forest types;

- classification and positional accuracy; and

- proprietary data formats.

The importance of national metadata standards was recognised at the outset of the programme, and ANZLIC guidelines for core metadata elements developed for use in Australia and New Zealand were adopted for all RFA data.

Many scientists have noted that electronic access to datasets is the most efficient way of distributing data to a wide variety of users. There are a number of examples where such practices are already commonplace, such as national statistical data. This has also been the case with natural resources data, and it has been argued that electronic access to natural history collections data will enable biodiversity to be mapped with other parameters, improving both science and environmental management (Winker 1999). However, there are problems related to the networking of data in this way, particularly with regards to ensuring quality of networked datasets. Winker observes that around 50% of the records of bird species in North America have yet to be computerised, and proofreading lags far behind. He explains that data quality is an issue when releasing error-ridden datasets to users who have little familiarity with the biases and errors in collections. Whilst the benefits of data sharing can clearly be seen, the scientific community must be responsible when releasing data that may be misinterpreted or that may have associated errors.

CHAPTER 2

DOCUMENTED DATA POLICIES

INTRODUCTION

This chapter presents the results of the work on reviewing the documented data policies of the organisations contacted during the study described in Chapter 1. The chapter is divided into five main categories of organisation, namely statistical institutes, mapping agencies, institutes for natural resources, environmental monitoring organisations and the Earth observation sector. For each category the discussion follows the six data policy themes running throughout this book, namely ownership, intellectual property rights, standards and metadata, licensing, pricing policy and archiving policy.

STATISTICAL INSTITUTES

Ownership, privacy and confidentiality

Most national statistical organisations claim ownership of the data and publications they produce, or share ownership with other organisations (such as universities and other public bodies) that may be responsible for the production of data. However, many recognise that statistics form an indispensable part of society's infrastructure and data is therefore often treated as a collective good to be disseminated as widely as possible. In Germany, for example, the Statistisches Bundesamt (Federal Statistical Office) equates basic statistical data with public roads; it is part of the state infrastructure that must be accessible to all.

Privacy is a key issue for national statistical organisations, and particular attention is paid to protecting the anonymity of individuals and businesses. In many cases, this protection is in the form of national and international legislation, such as the Austrian Federal Statistics Act 2000, the Irish Statistics Act 1993 and the Spanish Statistical Secrecy laws. Countries within Europe are also governed by European Union laws on statistical secrecy. In some cases, data on individuals is released, but this is under strictly controlled circumstances. In the UK, for example, data relating to individuals is only released when the National Statistics agency is satisfied that the data is to be used exclusively for justifiable research and that the information is not reasonably obtainable elsewhere.

International organisations, such as the World Trade Organization (WTO) or the World Health Organization (WHO), generally maintain ownership of data and publications they produce. International organisations take a similar approach to privacy as national statistical institutes, with protection of individuals a high priority. For organisations such as Eurostat, EU laws ensure privacy is respected.

Intellectual property rights and associated legal frameworks

Most statistical organisations maintain intellectual property rights (IPR) over their data, although the extent to which they are controlled varies between organisations. Liberal reuse and redistribution of data is often granted as long as the organisation responsible is quoted as the source. In some cases, such as the Irish Central Statistics Office and National Statistics UK, such copying and reproduction is only granted for non-commercial research purposes. In Germany, the Statistisches Bundesamt follows a three-tier structure to IPR depending upon which of the following three marketing models is used:

(1) The 'outer band' represents the basic information requirement of the general public that is free of charge. Data within this category can be reprinted and disseminated, providing the source is acknowledged.

(2) The 'middle band' represents the standard information requirement that can be attached to certain target groups. This data can also be reprinted and disseminated for non-commercial purposes, providing the source is acknowledged.

(3) The 'core area' consists of customer-specific information attached to special users. Products in the core area are subject to restrictions that may, where appropriate, be relaxed by negotiation.

In other cases, controls on IPR are more open. For example, Statistics Belgium has no explicit control mechanisms for IPR, primarily because it is a non-profit organisation and therefore has no particular financial incentive. Statistics Belgium therefore treats reactions to copyright infringement on an *ad hoc* basis.

International organisations also maintain IPR over their data. In many cases, this is not for financial gain, but so that the organisation can regulate the use of their data. For example, the WHO maintains IPR to ensure that information is used in accordance with the principles of the organisation.

Standards and metadata

Most statistical organisations follow standards laid down by international bodies such as the United Nations, the Organisation for Economic Co-operation and Development (OECD) and the European Union. This allows international comparability of statistical data, as well as allowing organisations to profit from work carried out by other statistical organisations. In the EU, most countries base their quality reports on the concepts of Eurostat, which define the quality of statistical products with reference to the following six criteria:

(1) relevance to statistical concepts;

(2) accuracy of estimates;

(3) timeliness in disseminating results;

(4) accessibility and clarity of information;

(5) comparability of statistics; and

(6) coherence

For certain fields, there is a legal obligation for statistical institutes to provide quality reports to Eurostat, such as external trade statistics, structural business statistics and for the survey on continuous vocational training.

UN and OECD recommendations are also followed by most statistical organisations, but some within the EU regard these as being of less priority than those of Eurostat. Another important data standard that many statistical organisations follow is the Special Data Dissemination Standard (SDDS) of the International Monetary Fund (IMF).

Metadata standards are regarded by many as being less defined than for the statistical data itself. Most statistical institutes have accompanying documentation with their data, but this does not necessarily follow international standards.

Licensing, distribution and dissemination

Many national statistical institutes regard their data as a 'public good' and therefore place as few restrictions on its use as possible. However, for some purposes (for example, for commercial use), institutes operate licences for the redistribution of data.

The extent of licensing activities varies between organisations. Some, such as Statistik Austria, do not have any licensing activities regarding their data. Others, such as National Statistics UK, require a licence agreement if the data is to be used for a purpose other than for private research or study. Licensing is also often used to control the use and distribution of more sensitive micro-level data that may be required for research or planning purposes, or for any data that is charged at the market price and not freely available. For many statistical institutes, the only requirement for redistribution or use of their data is that they are referenced as the source.

Most national statistical institutes aim to disseminate their data as widely as possible so that it can be obtained by anyone who wishes to use the data. For many statistical organisations, the Internet is an increasingly important tool for data dissemination. Many organisations make more general datasets available to download free of charge from websites. Statistical data is also disseminated through many other mediums in order to maximise its distribution and make it as accessible as possible. These include books and other publications, diskettes and CD-ROMs, emails and brochures, many of which are also made available electronically in Portable Document Format (PDF). More detailed datasets and tailor-made statistics are also provided by many statistical organisations. These are less widely disseminated and are generally only available on request.

International organisations also aim to disseminate their data as widely as possible, although many do not consider their data a 'public good'. For example, the WTO licenses some of its data collections exclusively to private sector distributors, who then make them available for a fee.

Pricing policy

General statistical datasets are provided free of charge via the Internet by most national statistical institutes, or at the marginal cost of production for books and publications. Notable exceptions to this rule include Statistisches Bundesamt, which follows a pricing policy according to the three classes of data defined in their marketing model (see above).

Pricing policy with regards to individual requests and other services varies more between organisations. Some seek to recover only the marginal cost of reproduction for such services, the majority of which represents staff time and material and printing costs. Others either charge the full market price or a standard fee for such services, which varies between organisations.

Tailor-made statistical data for research and educational purposes is often provided at a reduced rate by many organisations.

Archiving policy

The long-term value of national statistical data is generally recognised by most organisations. Consequently, statistical organisations archive their data for long periods of time, although the method and format by which data is stored is often variable. Recent data is generally held in electronic and hard-copy form, with older records more likely to be in hard-copy format only. The date from which records were first stored electronically varies between organisation.

Some organisations, such as the WHO, have begun scanning older and more valuable hard-copy documents and making them available electronically in PDF format.

Summary of the data policies of statistical institutes

Statistical institutes are often publicly-funded organisations, and the conditions regarding the use and redistribution of their statistical data are often unrestrictive. Whilst institutes typically maintain ownership of the data, the information is regarded as a public good that should be made available to all interested parties. Consequently, the only condition that is generally placed upon the use and distribution of statistical data is that the institute is referenced as the source. In many cases, data is available free of charge or at the marginal cost of reproduction, often via websites or in the form of publications. However, datasets that contain information relating to individuals or individual businesses are typically strictly controlled for reasons of privacy. In many cases, these rules are governed by international law. Similarly, many statistical institutes follow international regulations and standards (such as those defined by Eurostat and OECD) regarding the production of their data, so that statistical datasets are internationally comparable. Older statistical data is generally viewed to have a historical value, and archives of statistical information are therefore maintained over long periods.

MAPPING AGENCIES

Ownership, privacy and confidentiality

Ownership of map data is generally claimed by the agency responsible for its production or, in the case of some national mapping agencies, by their respective governments. Ownership rights on map data are rigorously upheld in most cases, primarily due to the amount of time and capital that is invested in the production of map data. There are a number of exceptions to this view, notably in countries such as Australia where spatial data collected by agencies in the public interest is viewed as a 'public good' and made widely available at marginal cost prices. There are also a number of examples where companies supply map data to which they do not have ownership rights. For example, Getmapping plc supplies a number of datasets that contain information produced by Ordnance Survey of the UK. In this case, Ordnance Survey maintains ownership of this data, but Getmapping are allowed to supply the information under a partnership agreement with Ordnance Survey.

Standard datasets held by mapping agencies are not generally affected by issues of privacy and confidentiality. However, geographic information produced for the Swedish territory cannot be stored in databases or disseminated before permission is obtained from the Military Relations Division of Lantmäteriet, the national mapping agency. Some countries also view map data as sensitive information that should not be available to the general public. There are therefore a number of national mapping organisations that restrict access to map data for reasons of national security.

Intellectual property rights and associated legal frameworks

IPR on map data generally remain vested in the agency responsible for its production or their respective governments. In most cases, the IPR on the data are tightly controlled, and any reproduction or redistribution of data is normally accompanied by a number of strict copyright agreements to prevent unlicensed distribution to third parties. Such tight restrictions on IPR are due mainly to the cost of production and the commercial value of map data.

Countries that follow a national infrastructure approach to spatial data, such as Canada, Australia and the US, generally view IPR issues differently. Copyright is not seen as a means to prevent data use, but is viewed as a method to protect data integrity and allow recognition of the quality of data. The Australian Commonwealth Office of Spatial Data Management, for example, does not place restrictions on the use of its data but does require the user to acknowledge Commonwealth copyright in the data.

Standards and metadata

Mapping agencies generally produce data that follows national rather than international standards, although this can vary with the type of information that is produced. For example, the National Survey and Cadastre of Denmark bases the cadastral register and topographic database on national standards, but nautical charts and the geodetic reference system follow international standards. In Australia, spatial data is also produced to national standards. Geographic Information System (GIS) data and information produced by mapping agencies are generally provided in the proprietary formats of major GIS vendors.

Most agencies already have, or are constructing, metadatabases describing their data. Most agencies follow international standards in the production of metadata, such as ISO 19115 for indexing metadata, although this is not always the case.

Licensing, distribution and dissemination

Most mapping agencies control the distribution and dissemination of their data through licensing procedures. Map data is increasingly being produced and supplied in digital format. Mapping agencies therefore offer different licences that are designed to meet the requirements of different users. For example, users who may simply wish to display a limited number of fixed map extracts on the Internet can use a different (and less expensive) licence than those who wish to take material and incorporate it into an application for onward sale and licensing. Licences for data that is to be exploited commercially are often treated separately, and in many cases mapping institutes collect royalties from the value of the final product.

Other agencies licence data according to the number of users who propose to use the data. For example, the National Land Survey (NLS) of Finland offers three types of licences: basic licence (1–5 users); extended licence (6–20 users); and institutional licence (over 20 users).

Countries that have a national spatial data infrastructure take a different approach towards the licensing of map data. For example, the Commonwealth Office of Spatial Data Management in Australia allows fundamental spatial data to be freely distributed, but a licence accompanies all datasets. This licence does not restrict use of the data, but sets out the rights and responsibilities of the data user and provider, and absolves the Commonwealth from any liability arising out of the subsequent use of the data or products developed from the data.

The majority of national mapping agencies distribute data through retail outlets in the form of paper maps and CD-ROM. Some, such as NLS of Finland and Ordnance Survey UK, also sell and disseminate digital data directly through their websites. Other mapping agencies, such as Bundesamt für Kartographie und Geodäsie (BKG) of Germany, are in the process of developing a capability for Internet distribution. There are also commercial companies, such as Getmapping plc and Multimap, who sell and distribute data exclusively through the Internet.

Pricing policy

A large number of European mapping agencies act in a similar way to commercial organisations, and sell their products and licences for a price that reflects the market value of the data. Many governments take the view that the user rather than the taxpayer should pay for map data. The cost of licences is often dependent on the proposed use of the data, the amount of data, the spatial coverage of the data and the number of users who plan to use the data. In contrast, spatial data in countries such as Australia and Canada is freely available at the marginal cost of reproduction.

Archiving policy

Older map data of national mapping agencies is generally held by the national archives of their respective countries or by the agencies themselves. The value of older aerial photographs is also being recognised by mapping agencies, some of which (Ordnance Survey of Ireland and Getmapping plc, for example) are scanning aerial photographs and making them available electronically.

Summary of the data policies of mapping agencies

The majority of national mapping agencies maintain ownership of the data they produce, and many enforce strict copyright controls to restrict the reuse and redistribution of map data. The distribution and dissemination of map data is often therefore controlled via licensing agreements, which are priced at market levels or according to the proposed use of the data. For example, many mapping agencies offer institutional licences that allow for extensive copying and redistribution of map data within a specific company or organisation. Such licences can cost a considerable amount of money. A number of notable exceptions to this pattern are Australia, Canada and the US, where the restrictions regarding map data are more relaxed. Spatial data in these countries is often available free of charge or at the marginal cost of reproduction. In these countries, spatial data is viewed as a 'public good', and the benefits of allowing free access to spatial data are viewed as being greater than restricting access and charging for data at the market price. Map data is often held by the national archives of individual countries or by the mapping agencies themselves.

INSTITUTES FOR NATURAL RESOURCES

Ownership, privacy and confidentiality

Those institutes of natural resources that are publicly-funded generally maintain ownership of the data they produce, although the data is usually made available in the public domain for any interested parties. Natural Resources Canada (NRC), for example, states that as a publicly-funded federal government department, those of its data that are generated through public funding are deemed to be publicly owned, with NRC as custodians on behalf of the Crown. Similarly, data held by Bundesamt für Naturschutz is covered by the German law on environmental information (*Umweltinformationsgesetz*), which states that all information concerning nature and environment in public authorities must be freely available and free of charge. Most publicly-funded European organisations follow policies similar to this, and are influenced by protocols such as the European Commission Directive on Freedom of Access to Environmental Information.

In cases where data is collected on behalf of an international organisation, ownership rights may differ. A good example is forest condition data collected for the Institute for World Forestry. In this case, forest administrators of countries participating in the International Co-operative Programme on Assessment and Monitoring of Air Pollution Effects on Forests (ICP Forests) exclusively collect forest condition data. Consequently, the legal owners of the data are the countries themselves, whose representatives are National Focal Centres established in capital cities.

In some cases, environmental datasets are not deemed to belong to those who have collected them, and ownership resides with the employers of such data collectors or those who have paid for the data collection. Some of these employers/funders allow those who have collected the data to maintain ownership for a limited time. For example, the Irish National Council for Forest Research and Development (COFORD) funds applied research and development with forestry companies. The ownership of this research and the results of the research reside with the company, but the research must be made available to COFORD for a period of three years after completion of the project.

Institutes for natural resources may restrict certain types of data in order to protect privacy and confidentiality. For example, confidential information supplied by survey respondents, or data regarding the emissions by private firms into waterways or the atmosphere is likely to be withheld. In other cases, datasets may be made anonymous so that information cannot be traced back to the individual level. For example, ICP Forests data held by the Institute of World Forestry is completely anonymous, so that users of the data cannot find out who owns a particular forest.

Intellectual property rights and associated legal frameworks

Institutes for natural resources generally retain IPR on the data, reports and surveys produced under their jurisdiction. However, these rights do not normally constrict the use and dissemination of data, most of which is freely available in the public domain. In some cases, organisations allow employees who have collected data for internal research purposes privileged access to, and use of, such data for a certain period of time. After this period the data must be made publicly available.

Standards and metadata

Institutes for natural resources represent a diverse group of organisations that produce a wide variety of data types. Unfortunately, there are no universal standards for natural resources data. The UK Forestry Commission follows the UK National Statistics code of practice and related protocols for any statistical data it produces. Organisations such as NRC and the US Natural Resources Conservation Service (NRCS) follow their own designated standards or those laid down by their umbrella organisations (for example, the Federal Geographic Data Committee (FGDC) in the US). Most reputable organisations involved in research try to ensure that their data is fully calibrated and quality controlled to ensure that their results are reliable.

Many publicly-funded institutes maintain metadata with any datasets they produce. For example, the German environmental authorities maintain a harmonised metadata model called *Umweltdatenkatalog* that forms part of the German Environmental Information Network (GEIN). Similarly, the Danish Forest and Nature Agency (SNS) and National Environment Research Institute (DMU) document their data through a national metadatabase that contains information such as quality status, structure, publication and accessibility.

Licensing, distribution and dissemination

Many publicly-funded natural resources institutes aim to make their data publicly available with as few restrictions as possible. This is partly due to recommendations and protocols such as the European Commission Directive on Freedom of Access to Environmental Information.

NRCS follows the guidelines of the FGDC, the inter-agency committee that promotes the co-ordinated development, utilisation, sharing and dissemination of geospatial data on a national basis. NRC follows a similar policy to that of the NRCS, although there may be limited access to data for finite periods, to allow those who collected them priority interpretation and submission for publication.

Some research organisations, such as the UK Natural Environment Research Council (NERC), operate more restrictive licensing controls on their data. This is primarily to ensure that data provided free of charge for *bona fide* research is not used for commercial purposes. Others, such as the Institute for World Forestry, make their data available with the provision that certain conditions are met. These may include specifying the purpose for which the data is to be used and agreeing not to redistribute the data.

Data produced by institutes for natural resources is increasingly being disseminated via the Internet. For example, the SNS, the UK Forestry Commission and the Bundesamt für Naturschutz in Germany all provide access to data via their websites. Many institutes also disseminate their data through annual and thematic reports and through publication in scientific journals, some of which are also available in electronic format through the Internet.

Pricing policy

A large proportion of institutes for natural resources supply their data free of charge or at the marginal cost of reproduction. In many cases, data is available to view free of charge via the Internet. Datasets and reports supplied in printed hard-copy format are generally priced to recover the cost of reproduction, as most natural resources agencies operate on a non-profit basis.

A few agencies supply some types of data at the market price. For example, the Bundesamt für Naturschutz supplies certain ready-made products, such as GIS data, at the market level. Products produced for a special purpose, or for a specific customer, are also supplied at the market price.

The NERC in the UK sells a licence to use its data at a price which reflects the value of the data and the cost it has incurred in data acquisition. This policy operates for wholly commercial applications; data for educational or research purposes is provided free of charge or at a discounted rate.

Archiving policy

The archiving policies of natural resources institutes are variable and are normally dependent on the perceived long-term usefulness of individual datasets. Some organisations, such as COFORD for example, archive all their completed projects. Others periodically review the cost and benefits of continuing to maintain datasets, and then decide whether to continue to maintain or destroy the data. Some, such as NERC in the UK, publicly announce any intention to destroy data, allowing time for a response from interested parties.

Summary of the data policies of institutes for natural resources

Ownership of natural resources data is generally maintained by the organisations responsible for its collection, although the information is also normally placed in the public domain, particularly if the data is produced using public funding. In some cases, free access to this information is governed by a legal protocol, such as the European Commission Directive on Freedom of Access to Environmental Information. IPR regarding natural resources data are consequently unrestrictive, although the creators of data are sometimes given privileged access for a certain time to allow for the production of reports and publications. Natural resources information covers a wide variety of organisations and projects, and there are no universal standards for natural resources data. The archiving of natural resources data is variable between institutions, and often depends on the perceived long-term value of the data in question.

ENVIRONMENTAL MONITORING ORGANISATIONS

Ownership, privacy and confidentiality

The topic of environmental monitoring covers a large spectrum of organisations that often have a variety of policies and practices. Meteorological organisations, for example, generally maintain ownership of data and weather information they produce. However, such policies are heavily influenced by international agreements and institutions such as the World Meteorological Organization (WMO) and ECOMET.

Organisations responsible for monitoring environmental degradation and pollution also maintain ownership of their datasets, although public access to such data is often influenced by legal frameworks such as the European Commission Directive on the Freedom of Access to Environmental Information, the Aarhus Convention and national laws regarding freedom of information. For example, information and data produced using public funds is deemed to be publicly-owned in the US. Issues of privacy and confidentiality tend to arise in environmental data that may contain sensitive or confidential information on members of the public, companies or products. Data regarding threatened species (such as the location of nesting areas) is often made confidential. In a few cases, data is classified as confidential for reasons of national security. For example, environmental information regarding nuclear installations is restricted in the UK under the Anti-Terrorism, Crime and Security Act 2001. Similarly, the US National Oceanic and Atmospheric Administration (NOAA) may restrict certain foreign data and information according to international agreements, as well as any homeland security related issues.

Intellectual property rights and associated legal frameworks

IPR on meteorological data are generally held by the national meteorological agencies that produce the data. In the case of international bodies such as the European Centre for Medium-Range Weather Forecasting (ECMWF), IPR are also held by the organisation, except in the case of value added services generated by national meteorological agencies. In these cases, IPR are either shared between the service provider generating the value added service and ECMWF, or wholly owned by the service provider.

Most environmental research and monitoring agencies also maintain the IPR to their data, even though many place their data in the public domain. IPR are therefore not necessarily used to constrain access to data, but organisations require an acknowledgment if their data is used in further research or publications. IPR are also maintained to prevent third parties repackaging and selling the information and claiming rights over the original data.

Standards and metadata

The majority of meteorological agencies follow international standards recommended by bodies such as the WMO, ECOMET and ECMWF. Such policies ensure that weather observations and products are standardised, which is essential in order that global weather can be effectively monitored. Similar policies are also followed by other international monitoring agencies, such as the Intergovernmental Oceanographic Commission (IOC).

The importance of metadata is also recognised by environmental organisations, and the majority maintain some type of metadata system. Such metadata does not always meet international standards, although international bodies such as the European Environment Agency (EEA) have to meet the metadata standards of ISO 19115. The NOAA recommends that metadata be written for a user 20 years in the future: this ensures that a potential user will be able to find out all the required details about individual datasets, and also proves useful if the data needs to be re-used or corrected.

Licensing, distribution and dissemination

Licensing and distribution of meteorological data and products is heavily influenced by international agreements and organisations, most notably the WMO. Resolution 40 (see below) is one of the key agreements of the organisation, and requires that its members provide, on a free and unrestricted basis, 'essential data and products required to describe and forecast weather and climate, and to support WMO programmes'. WMO members must also provide selected data to the research and education communities on a free and unrestricted basis. In Europe, many countries follow a standardised policy of licensing and distribution for meteorological data and products, defined under the terms of ECOMET. Similar policies are also followed by the IOC, whose Member States are required to provide 'timely, free and unrestricted access to all data, associated metadata and products generated under the auspices of IOC programmes'.

Environmental institutes that are publicly-funded and act in the public interest generally make their information and data as widely available as possible. Many such organisations make their data freely available to interested parties with little or no licensing restrictions. However, a number of environmental organisations use licences not to restrict the use of the data but to ensure that attribution is given to the supplier organisation and that the copyright of the data is maintained. This data is increasingly being disseminated through the Internet in order to maximise its use and distribution. In cases where data is not available on the Internet, data is often disseminated in the form of publications and reports, or is obtainable by request from the organisation.

Pricing policy

The price of meteorological data is influenced by international organisations and agreements. The WMO regards meteorological data as essential to the wellbeing of humanity, and much of the data should therefore be exchanged on a free and unrestricted basis. Essentially, this means that basic meteorological information such as warnings, forecasts and observational data to a level declared to be in the public interest are made freely available. WMO Resolution 40 also ensures that some meteorological data and products are provided free of charge to non-commercial research and educational users.

The majority of meteorological agencies also produce value added services and data for commercial purposes that is made available at the market price. In Europe, prices for commercial use of data follow ECOMET guidelines. This price comprises the following components:

- an information price, which is the same for each National Meteorological Service (NMS);

- a delivery price, which is determined individually by an NMS; and

- the transmission price, being the cost of telephone, fax or other means of delivery.

Environmental monitoring organisations are often publicly-funded and not for profit, and most data is therefore provided at no cost through media such as the Internet. Where data is not available through the Internet, or where special requests are made, data is typically provided at the marginal cost of reproduction.

Archiving policy

Meteorological data is recognised as being a valuable historical and climate record, and most agencies have therefore compiled archives over many years. This has been in digital format over recent years, although many meteorological organisations hold records and information in hard-copy format dating back several decades. For example, the Technical Archive of the UK Meteorological Office holdings include:

- registers of meteorological observations – daily records of weather from many locations;

- working charts from the National Meteorological Centre;

- climatological returns dating back to 1855, covering temperature, wind, rainfall, solar radiation, snow and sunshine;

- autographic records; and

- upper-air data from radiosonde and pilot balloon ascents.

The archive also includes other documents such as diaries, papers and pictures relating to meteorology and the Meteorological Office.

The archiving of other types of environmental monitoring data often varies between organisations. The Finnish Environmental Administration, for example, has stored environmental data since the early 1970s, although they hold data that dates as far back as the 1850s. Others generally have policies that aim to reduce the holding of unnecessary information whilst maintaining those of permanent historical value. For example, if the user community does not require access to the data and resources to properly archive and service data are lacking, NOAA may purge its data holdings in accordance with the appropriate data management plan, not on a unilateral basis.

Summary of the data policies of environmental monitoring organisations

Environmental monitoring covers a wide spectrum of organisations in which there are a variety of policies and practices. Organisations responsible for monitoring environmental degradation and pollution generally maintain ownership of their data, although this information is also made freely available to the public. Similarly, meteorological agencies maintain ownership of their data, but this is heavily influenced by international agreements and organisations such as the WMO and ECOMET. IPR are generally unrestrictive for general environmental data and 'essential' meteorological data. However, more restrictive conditions are maintained on other meteorological data, such as value added products produced by individual meteorological agencies. With regards to standards, meteorological agencies follow those laid down by international bodies and agreements, such as the WMO. Other forms of environmental information cover a variety of datasets and sources, in which the use of standards is variable. However, most environmental monitoring organisations aim to meet international standards and agreements where they exist. Environmental data and research is often widely published and freely distributed, particularly if it is produced using public funding. Long-term archiving of meteorological data is commonplace, as the need for detailed and long-term climate records is well-recognised. Archiving practices for other types of environmental data are more variable, and are generally dependent on the perceived long-term usefulness of individual datasets.

EARTH OBSERVATION ORGANISATIONS

Ownership, privacy and confidentiality

Ownership of Earth observation data is maintained and often strictly upheld by the companies and organisations that supply the data. In the majority of cases, licences to use the data are sold whilst the ownership of the data is retained by the supplier. This ownership extends to copies of the data, regardless of the form or media on which the original and copies may exist. In cases where Earth observation missions are publicly-funded by bodies such as NASA and NOAA, ownership of data is maintained by the government and viewed as a 'public good'. In the case of the US, this data is often made freely available under the Freedom of Information Act.

Privacy and confidentiality regarding Earth observation data has become a more important issue as the spatial resolution of satellite data has improved. DigitalGlobe and Space Imaging, which operate the high resolution Quickbird and Ikonos satellites respectively, are subject to scrutiny by the US government to ensure that sensitive information is not released to unauthorised parties. There are a series of policies in place that allows the US government to restrict the distribution of imagery where national security is at risk or where US assets overseas may be placed at risk. In addition, it is prohibited to release imagery within 24 hours of collection so that US forces can exercise operational security checks if necessary.

Intellectual property rights and associated legal frameworks

IPR and copyright restrictions of Earth observation data are generally strictly upheld by the data suppliers. In cases where data or a derived product is placed in the public domain, proper copyright must be conspicuously marked and the data must not be redistributed without permission. Such copyright restrictions generally arise from the commercial value of the data, which the data supplier organisations could potentially lose through unauthorised distribution. Reproduction and redistribution of data are therefore only permitted within the organisation of the data use and according to the terms set out in licensing agreements.

Earth observation data recorded primarily for scientific research, rather than commercial purposes, generally have more relaxed policies regarding IPR. For example, some datasets produced by US federal agencies such as NASA and NOAA can be acquired and redistributed with few or no restrictions, as long as the user correctly acknowledges the data source.

Standards and metadata

The standards regarding Earth observation data depend upon the platform from which the data originates and the organisation responsible for supplying the data. For example, Radarsat International products are calibrated during processing to provide both geometric and radiometric corrections to the data and to ensure compliance with Radarsat's image quality specifications. However, data processed at other Radarsat network stations may not be calibrated to the same standards, which are defined by levels of certification. The EU Forum on Earth Observation (EUFOREO) project (Cannizzaro 2003) noted that metadata for Earth observation products are seen to be poor in many cases, with problems related to both the discovery of metadata needed to carry out an initial assessment of Earth observation data, and the access metadata needed to actually acquire and use them.

A number of organisations have cited problems with international standards that may hinder their adoption amongst data providers. International metadata standards allow catalogue interoperability but can be restrictive in terms of the fields and concepts they describe. For example, the Earth Observation Data Centre (EODC) in the UK holds many terabytes of Earth observation data and produces its metadata according to the FGDC standard. Part of its data holdings consists of data from the Along Track Scanning Radiometer (ATSR) instrument, but many ATSR-specific fields are not included in the FGDC standard. The metadata can be extended to include these fields, but this implies that the information is no longer 'standard'. Consequently, data providers are often tempted to abandon recognised metadata standards in favour of their own systems that are viewed as being more descriptive of specific datasets. The use of metadata standards is also hindered by the fact that many such standards are complex and detailed, and are supplied with many pages of documentation. Scientists who use Earth observation data are not necessarily remote sensing experts, and require simple metadata to assess the datasets' suitability for their purpose without complicated information that may not be relevant for their needs. Many data producers are consequently discouraged from attempting to use complicated standards, and document their datasets according to their own metadata systems. Another issue is that International Standards Organisation (ISO) metadata standards and documentation are not freely available and can only be obtained through the ISO for a fee. This is so that the ISO can meet its operational costs, but this may discourage data producers from adopting ISO standards when their own system for metadata can be used immediately and free of charge.

Data from airborne platforms, as opposed to satellites, often requires more rigorous quality controls and standards due to the variability in factors such as the pitch and yaw of the aircraft. For example, the NERC Airborne Remote Sensing Facility in the UK performs quality control checks throughout all stages of its operations. This process includes the following steps:

- sensor calibrations performed before, during and at the end of the operational season;

- in flight, the instrument operator checks the correctness of the sensor parameters in real time;

- at landing the sensor and navigation datasets are pre-processed and checked for anomalies;

- data transcriptions are checked for integrity, and photography for cloudiness and illumination indexes; and

- processed data is checked for accuracy of geometric correction and geo-location.

Licensing, distribution and dissemination

The use of Earth observation data is generally governed by the use of licences that control the distribution and dissemination of data. Data suppliers that operate on a wholly commercial basis, such as Space Imaging and DigitalGlobe, generally have more restrictive licensing arrangements regarding the use of their data. By comparison, data that is recorded for scientific and research purposes by organisations such as NASA is often freely available with fewer restrictions.

DigitalGlobe and Space Imaging offer licences for the use of their data which generally vary according to the number of people who propose to use the data (single or multiple organisation licences, for example). These licences offer varying standards of flexibility in image sharing and redistribution.

The conditions attached to Earth observation data supplied by the European Space Agency (ESA) depend on the use of the data, which falls into two categories: research and applications development; and operational and commercial use. Data for research and applications development is supplied directly by ESA from its own facilities, whilst data for commercial use is supplied by distributing entities selected by ESA. These distributing entities then establish their own data distribution scheme by defining prices and negotiating contracts and sub-licences.

Some Earth observation data is provided with very few restrictions and licensing conditions. For example, NASA follows a policy of full and open sharing of Earth science data obtained from US government funding as soon as data becomes available. Products such as QuikSCAT data and data from the Moderate Resolution Imaging Spectroradiometer (MODIS) instrument are provided free of charge for scientific and educational use, with virtually no restrictions. The only conditions that govern the use of such data are that it is not redistributed for profit and that appropriate acknowledgments are made in any resulting publications.

Pricing policy

The price of Earth observation data is often dependent upon whether the data in question is to be used for commercial or research purposes. For example, ESA supplies European Remote Sensing Satellite (ERS) and European environmental satellite (Envisat) data at or near the cost of reproduction for research purposes that meet its mission objectives. For commercial purposes, data is supplied at a market price that is determined by the various distributing entities. Other publicly-funded organisations such as NASA offer a large amount of data free of charge via the Internet, with little or no restrictions regarding its use.

Data supplier organisations that operate on a wholly commercial basis, such as Space Imaging and DigitalGlobe, supply their data at the market price. This price can often be quite substantial (around US$6,000 per scene for Quickbird imagery), which reflects the high spatial resolution and accuracy of such data.

European Organisation for the Exploitation of Meteorological Satellites (EUMETSAT) data is supplied to the NMS of Member States free of charge. For non-Member countries, the following pricing policy applies (EUMETSAT 2004):

- Data is supplied free of charge for official duty use by NMSs of countries with a gross national income per capita below US$3,500.

- For countries with a gross national income per capita higher than US$3,500, data for official duty use is supplied according to a table of fees based upon the quality and frequency of the data.

Data for commercial and other purposes is charged at a market price that is reviewed by the EUMETSAT council at regular intervals.

Archiving policy

Most organisations involved in Earth observation archive their data, and the nature of the datasets means that this is in digital form. The volume of archived remote sensing data can often be extremely high for some organisations. For example, the German Remote Sensing Data Center (DFD – Deutsches Fernerkundungsdatenzentrum) estimates that their archive will grow to hold between 100 to 300 terabytes of data by 2005, with 30 gigabytes of data currently being added every day. In the case of the NASA Earth Science Enterprise, each Earth science dataset is subject to peer review to determine its merit for long-term archiving. The peer review process considers for long-term archiving:

- all data acquired by or in support of Earth Science Enterprise funded research projects;

- all data acquired systematically by Earth Science Enterprise funded missions for the purpose of documenting long-term environmental variability; and

- all accessible data concerning natural disasters and other extraordinary events identified by the Earth Science Enterprise.

The accessibility of archived data often depends on the length of the mission and the age of the data. For example, Ikonos was launched in 1999 and Space Imaging maintains an extensive archive of data since this time. By comparison, EUMETSAT holds data covering a time span of over 20 years, during which the technology and media on which data is stored has been significantly updated. For example, data recorded since December 1995 is stored on digital linear tape. This is a different format to that used before December 1995, and the same retrieval mechanism cannot be used for data from before this date. A project to transfer the old data onto the new format is currently underway, but will take some years to complete.

Summary of the data policies of Earth observation organisations

Ownership of Earth observation data is generally strictly upheld by the respective organisations involved. In many cases, this ownership extends to copies of the data, regardless of the forms or media in which the copies or original may exist. Redistribution of Earth observation data is therefore strictly controlled by copyright agreements in many cases, ensuring that the supplier organisations do not lose out through the unauthorised redistribution of data. Commercial Earth observation organisations generally sell licences to use their data, not the actual datasets themselves, in order to control the distribution of their products. In such cases, the cost of the licences depends upon the number of data users and the proposed use of the data. Earth observation data that is recorded primarily for scientific research is generally made available with fewer restrictions. Data from a number of NASA missions for example, is made freely available via the Internet without the use of licences. Most Earth observation missions archive their data on a long-term basis, but access to older data can sometimes be difficult due to outdated storage media or reading technology.

CHAPTER 3

SOCIO-ECONOMIC DATA

INTRODUCTION

In addition to the environmental data that will be needed for global environmental monitoring, socio-economic data will also play a key role. Such information is useful for environmental studies, for example, in estimating the degree to which population and urbanisation may affect a particular habitat. Socio-economic data is also likely to play an important role in the security aspect of Global Monitoring for Environment and Security (GMES), and provide the necessary demographic information useful for directing aid as well as investigating infrastructure and amenities for disaster relief. A large amount of socio-economic data is produced and disseminated by national statistical agencies of individual countries, plus international agencies such as Eurostat, the UN and the World Trade Organization (WTO). A selection of the themes relevant to global environmental monitoring is discussed in this chapter, namely agricultural productivity, health, tourism, trade and transport. A separate section deals with population census data.

THEMES

Agricultural productivity

A wide variety of agricultural statistics are available from Eurostat's website (Eurostat 2004), and include themes such as agricultural price trends within the EU, livestock population and production volumes of agricultural crops. Free data that can be downloaded from the Internet include monthly averages of collected cow's milk products, pork, beef and veal slaughtering, as well as various structural indicators.

Data relating to agricultural productivity is also produced and distributed by the Food and Agriculture Organisation (FAO) of the United Nations via the online resource known as FAOSTAT (FAOSTAT 2004). This resource contains over one million records covering international statistics on production, trade, fertilisers and pesticides, land use and irrigation, forest products, fishery products, population, agricultural machinery and food-aid shipments. The information is presented as a searchable database in which the user can search for particular records according to a number of parameters. Whilst this method is useful for locating specific information about a named country, for example, it does not allow the downloading of the entire dataset and detailed analysis of the data is therefore limited.

Data relating to agricultural productivity can also be obtained from national agencies, for example the Department for Environment, Food and Rural Affairs (DEFRA) in the UK or the Department of Agriculture and Food in Ireland. In the case of the UK, a large amount of this statistical information is produced in association with the national statistical organisation and is available to download from the DEFRA website (DEFRA 2004). The DEFRA statistical datasets are grouped according to topics such as agriculture and food, farming statistics, environment, fisheries and rural affairs. Within these topics, different data products can be accessed, including economic reports, publications and statistical data. The Irish Department of Agriculture and Food follows a similar model, with a number of summarised statistics available to view on the website, as well as publications, reports and other information. A similar situation exists in the US, and a large amount of agricultural data is provided free of charge via the website of the Economic Research Service of the US Department of Agriculture (ERS 2004). A total of 39 different tables of data are available, which cover themes such as livestock, crops, agricultural trade and prices. The information can be downloaded as Excel spreadsheets or in Portable Document Format (PDF) with few or no restrictions.

Access to data relating to agricultural productivity is often produced by, or in association with, national statistical organisations. Agricultural data is often therefore treated in the same way as other statistical information, and is made freely available to members of the general public.

Health and healthcare

The World Health Organization (WHO) is an obvious source for data relating to human health and healthcare. A large amount of data from WHO technical programmes is available via the online WHO Statistical Information System (WHO 2004b). Datasets available include the WHO mortality database, immunisation statistics, burden of disease statistics and population statistics. Users can freely download the majority of this data.

Some of the datasets are also provided in a range of formats, which are designed to suit users with different purposes and experience. For example, detailed raw datasets held in the WHO Mortality Database can be downloaded for those with sufficient experience and who wish to perform detailed analyses on the data. The data is also available in summary tables that are suitable for novice users or those who need the data for straightforward reference purposes.

In addition to statistical data, the WHO also provides access to a limited number of spatial datasets through its website. For example, the WHO Global Atlas of Infectious Diseases (WHO 2004a) allows users to query and search for specific information held in the database, and outputs the data in reports, charts and maps. The database provides a mapping interface that allows users to select specific geographical areas and create maps of diseases, location of health facilities, schools and roads, and other geographic features. Static maps and related documents are also available to download from the website.

Statistical indicators on health are available from the United Nations Statistics Division (UNSD). Tables detailing life expectancy and infant mortality for every country are freely available to view (although not to download) on the website. The data is also provided with detailed explanations and technical notes about how the data was recorded and its reliability.

National statistical organisations are a reliable source for health and healthcare related data for individual countries. Much of this data is provided to the public free of charge, as many publicly-funded statistical institutes view their products and data as public goods. However, reliable statistical information from national statistical institutes is generally not as widely available for many less economically developed countries. For example, the communications and computing infrastructure in many African nations is not as widely developed as that of developed nations, and therefore providing access to statistics via the Internet is not possible or not appropriate.

Industrial productivity and trade

Like agricultural data, industry and trade statistics are often produced and/or disseminated by national statistical organisations. For example, key indicators for industry and trade (such as the index of turnover, producer price index and industrial production) are available to download for free from the Eurostat website (Eurostat 2004). However, these datasets are fairly generalised and do not give much detailed information. More detailed datasets, or tailor-made datasets, have to be ordered in advance from Eurostat, for which a charge is levied.

Datasets relating to individual countries can be obtained from the relevant national statistical organisation, and in most cases this is free of charge and available on the Internet. In some cases, these datasets are provided as simple tables presented on a web page, whilst others are more detailed. For example, Statistics Denmark allows users to select data from a dataset according to a number of parameters, including region, assessment and time period: an example is given in Figure 3.1, overleaf.

Data and statistics relating to international trade can be obtained from international organisations such as the WTO. International trade statistics can be downloaded directly from the WTO website (WTO 2004), covering selected long-term trends, trade by region and trade by sector. These statistics are also available in paper format or CD-ROM, and can be ordered through the WTO bookshop. Files can be downloaded in a variety of formats, such as MS Excel or PDF format, and can also be viewed online as tables.

Data produced by national statistical organisations generally meet international standards so that information is internationally comparable. Most data relating to industrial productivity follows the standards recommended by international bodies such as the United Nations, the Organisation for Economic Co-operation and Development (OECD), and Eurostat.

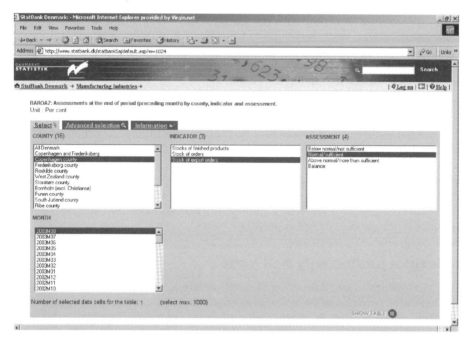

Figure 3.1: The Statistics Denmark web page allows the user to select industry statistics according to a number of parameters. Source: www.statbank.dk

Leisure and tourism activity

Data and statistics relating to leisure and tourism can also be obtained from many national statistical organisations for free. For example, in the UK, users can download statistics on business tourism, domestic and overseas tourism and visitor expenditure in the UK and overseas. Similarly, the Federal Statistical Office in Germany provides statistical data including employment and turnover in the tourism industry, the number of arrivals from abroad and city tourism.

Data relating to tourism can also be obtained from international organisations such as the World Tourism Organisation. This organisation provides statistical information on tourist arrivals, methods of transport and international tourism expenditure for more than 200 countries from 1985 to 1999. However, this data is not provided free of charge and users must register to gain access to the database. The cost of the data is charged per data cell, at a rate of US$0.05 for affiliate users and US$0.10 for other users. Eurostat also provides information and data relating to leisure and tourism that can be obtained online. Some of this information can be downloaded free of charge, but some datasets must be paid for: for example, the *Tourism Statistics Yearbook* for 2003 can be ordered on CD-ROM for €120.

Transport infrastructure

Other socio-economic data, such as transport infrastructure, is available through organisations such as the European Environment Agency (EEA). The EEA website (EEA 2004) provides fact sheets that include graphs, statistics and analysis on the capacity of transport infrastructure networks in the EU and accession countries. Datasets detailing the number of cars, two-wheeled vehicles and ships within the EU is also provided by the EEA. Infrastructure data is also available from the Energy and Transport directorates of the European Commission. This data is free to access and is displayed as tables on the EUROPA website (EUROPA 2004).

The World Bank provides datasets that compare the transport infrastructures of a number of countries. This includes the Railways Database, which provides information on scale, output and performance for over 90 railways worldwide. The dataset can be downloaded in its entirety in MS Excel format or, for those users who wish to access the underlying data, it can be downloaded on a country by country basis. Information describing the dataset can be downloaded from the website, and includes summary tables for all the railways covered.

For individual countries, national statistical organisations represent the best source of data relating to transportation, for example the nature of the goods transported, the number of vehicles on the roads and car ownership per head of population. At the European level, some of these datasets can be obtained from Eurostat: more generalised datasets can be viewed free of charge from the Eurostat website, but a charge is levied for more detailed information.

EUROPEAN POPULATION CENSUSES

Core data

Census data represents one of the most important sources of socio-economic data, supplying information on numbers of people, where they live and their associated needs. Most countries conduct a population census, although the scope and nature of the questions can vary between countries, as can the degree to which this information is made available to the public. Almost all countries that carry out a census have a privacy policy regarding the personal information of participants, and data relating to individuals is almost always anonymised or restricted.

The United Nations Economic Commission for Europe (UNECE) and Eurostat produced a set of recommendations for the conducting of censuses within the European Community region in 2000. The aims and objectives of these recommendations were:

• to provide guidance and assistance to UNECE Member States and other interested countries in planning their censuses; and

• to facilitate and improve international comparability through the harmonisation of data, definitions and classifications of topics.

Country examples

This section summarises population census data availability through the web for a sample of countries in Europe plus the US. The websites for the following countries were used to collect information on population censuses:

Croatia	www.dzs.hr/Eng/Census/census2001.htm
Estonia	www.stat.ee/frames.aw/section=617/set_lang_id=2
Finland	www.stat.fi/tk/he/vaestolaskenta/vaestolask_en.html
France	www.insee.fr/en/recensement/page_accueil_rp.htm
Ireland	www.cso.ie
Spain	www.ine.es/inebase
Switzerland	www.bfs.admin.ch/content/bfs/portal/de/index.html
UK	www.statistics.gov.uk/census2001/default.asp
US	www.census.gov/main/www/cen2000.html

Croatia

The Census of Croatia is produced following the recommendations of Eurostat and the UNECE. All core topics described by the recommendations are included within the census, as well as 50% of the non-core topics. Tables containing the results of the census data are available from the website, which includes data from both national and regional levels. A limited number of graphs are also available.

Estonia

Results from the 2000 Census can be downloaded from the website of the Statistical Office of Estonia and cover core themes such as age, ethnicity and education. In addition, the user can choose to download complete datasets, or data according to a particular parameter (for example, gender or level of education).

Finland

Tables and publications based on census data are available in digital format for free from Statistics Finland's self-service system StatFin. Chargeable print-outs and tables can be ordered by email or telephone. Special compilations of data can also be ordered by the user for a fee. Statistics Finland produces a range of CDs from census data, including data in map form and by postcode areas.

France

The most recent census in France was held in March 1999. The results are made available by the national statistical institute INSEE and can be freely downloaded from the website. Users can select a particular theme, region and the type of product that they wish to obtain, and the available data is then displayed. Whilst most key data is available to download free of charge, databases are priced at €11.50 and are provided with the relevant software so that the data can be analysed.

Ireland

Early results from the preliminary report of the 2002 census are available for free from the Irish census website. Definitive population figures that include factors such as age, marital status and occupation are made freely available on the website. Thematic maps and detailed commentary are also available.

Spain

The most recent Spanish census took place in 2001, the results of which are distributed by the Spanish National Statistical Office. Much of this data is distributed via the website, which provides data according to census parameters such as sex and age group. Individual parameters can be selected via drop down menus on the website, and the requested data is then presented on the screen. Maps displaying census data at the municipal level can also be downloaded from the website, along with the required software tools to view and analyse the data. All data available through the website is provided free of charge.

Switzerland

The census in Switzerland has been held every ten years since 1850, the most recent taking place on 5 December 2000. The Swiss Federal Council specifies that the results of the census are made available to any interested parties, including specialists in research, industry and politics as well as the general public. The Swiss Federal Statistics Office publishes the results of the census in a variety of formats including the Internet, and both hard-copy and electronic versions of reports and summaries. Swiss census information is provided free of charge.

UK

The census in the UK has been held every 10 years since 1801, the most recent being in 2001. The results of the census are freely available online from the National Statistics website at local, regional and national levels. Thematic maps covering a large number of themes (health, employment, ethnicity, etc) are also free to download in PDF format. Other derived census products, such as national and regional rankings by theme and other commentaries (for example, differences in population estimates, response rates, etc) are also available to download free of charge.

US

The US Census Bureau conducted its most recent census on 1 April 2000. Comprehensive access to information, maps, databases and other information derived from the census is provided online by a dedicated data portal called the Census 2000 Gateway (see Figure 3.2 overleaf).

The site provides a very large amount of census information to any interested users, either free of charge or at marginal cost, and in a variety of media and formats. The website enables the data to be manipulated and presented in a

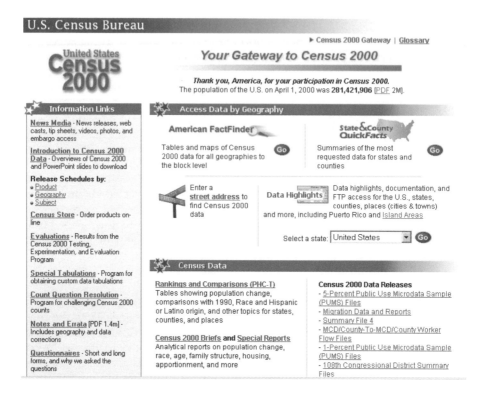

Figure 3.2: The US Census 2000 Gateway web page. Source:
www.census.gov/main/www/cen2000.html

variety of ways. For example, American FactFinder is an interactive thematic map
that can provide data on any of the census themes from scales that range from
national to block level (see Figure 3.3). Other options include a function that
allows users to search for data according to a particular address and File Transfer
Protocol (FTP) access to datasets from local, regional and national levels.

CONCLUSIONS

Socio-economic data is often created by publicly-financed statistical institutes,
and is therefore often distributed to the public on a free and unrestricted basis.
International comparability is an important factor for national statistical
organisations, and therefore most of the datasets are produced to internationally
recognised standards, such as those of the UN and Eurostat. The amount of
information that is provided by the various institutes often differs between
organisations. Many only release less detailed datasets of key variables and then
supply more detailed datasets on demand. For example, in some cases, statistics

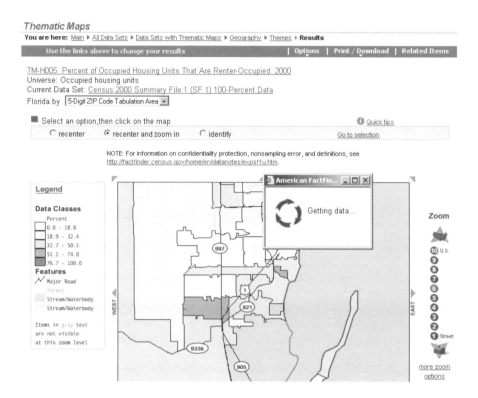

Figure 3.3: The US Census Bureau data portal known as American FactFinder, which allows users to browse census information from national to street level. Source: www.census.gov

relating to industrial production can only be viewed as a table displayed on the Internet, whilst other organisations allow entire datasets to be downloaded for free.

Similarly, whilst most countries provide free access to census results, the amount of detailed information that is easily accessible differs considerably between countries. For example, some countries only provide tables of the key census variables, in which the data cannot be downloaded or easily manipulated. Many European countries, such as the UK, France and Finland, have more detailed census results available on their Internet sites, and are often accompanied by thematic maps, graphs and detailed analysis. The US Census Bureau provides the most detailed and indepth census results online, with a dedicated data portal that allows downloading of datasets and publications as well as interactive maps and searchable functions.

It is clear that whilst most countries are keen to distribute socio-economic data and statistics as widely as possible, the amount of information that is easily available to users can vary considerably between countries and organisations.

CHAPTER 4

EVIDENCE OF DATA ACCESS CHALLENGES

INTRODUCTION

This chapter presents evidence from the practical experience of applications projects using data policies in action. Members of 14 projects funded by the European Commission and by the European Space Agency (ESA) were interviewed to elicit their practical experiences with access to environmental data and to discuss their views on data policy. In addition, the discussions with members of the projects allowed the inclusion of evidence from wider experiences than just the projects themselves. The relevant projects are listed in Table 4.1 overleaf, and consisted of eight EC Global Monitoring for Environment and Security (GMES) projects interviewed in face-to-face meetings, three EC GMES projects that provided written input and three ESA TESEO (Treaty Enforcement Services using Earth Observation) projects interviewed in a round table meeting. Information on data policy experiences from other projects is included where appropriate.

This chapter uses the common approach of the six main data policy characteristics to provide its structure, namely ownership, intellectual property rights, standards and metadata, licensing, pricing policy and archiving policy.

OWNERSHIP, PRIVACY AND CONFIDENTIALITY

Where an organisation is responsible for capturing its own data, then ownership is clear, for example, with laboratories that measure methane concentration in air samples. A similar situation can be claimed for aerial photography when an organisation commissions and pays for an air photography campaign, and then owns the photographs or digital scanner data. In the OCEANIDES project at the European Commission's Joint Research Centre (JRC), the ownership of the data produced by the project itself lies with the project members, although they do note that the data cannot be used for commercial purposes. The GMES-GATO (Global Atmospheric Observations) project notes that data from ground-based observations are typically owned by one person or a small group, but this can lead to access problems because that person or group is often not equipped to make their data more widely available. Privacy and confidentiality restrictions have been used by publicly-funded authorities to refuse data to the EUROSION project. This was particularly the case for ground-based measurements such as erosion rates, beach profiles, geomorphology data and sediment drift characteristics. There is no mechanism in Europe to force publicly-funded data to be made available to the wider user community.

EC GMES projects: face-to-face interviews	EC GMES projects: written comments	ESA TESEO projects: round table discussions
BIOPRESS: pressure on biodiversity	ESONET: seafloor observatory network	Carbon monitoring and the Kyoto Protocol
DISMAR: marine pollution and water quality	EUROSION: coastal erosion management	Desertification assessment
EOLES: disaster management	GMES-GATO: global atmospheric observations	Ramsar Convention on Wetlands
GMES-RUSSIA: network for GMES in Russia		
LADAMER: land degradation in Mediterranean Europe		
MERSEA: marine environment and security		
Meth-MonitEUr: atmospheric methane monitoring		
Siberia II: carbon accounting for Northern Eurasia		

Table 4.1: The projects examined for evidence of data access challenges. All the discussions and information collection took place during 2003. TESEO is an ESA action on environmental treaty enforcement using Earth observation data

Some difficulties arise when projects use environmental data captured by others, which is commonly the case in projects that use Earth observation data. The owners of Earth observation data, for example ESA or the Centre National d'Etudes Spatiales (CNES), grant a licence to use the data and do not convey ownership of the data to the purchaser. A clear example of this is in the licence agreement for access to the Shuttle Radar Topography Mission (SRTM) data available from the German Aerospace Centre DLR (DLR 2004):

1   The SRTM/X-SAR products are protected by international copyright laws.

2   The license includes the delivery of height and image data as well as the medium of delivery and permits use of one copy only of the product. The copyright and the data themselves remain the property of DLR.

The Co-ordination of Information on the Environment (CORINE) land cover dataset is a good example of uniform ownership and consequent access. For the highest level of data (25 ha polygons), the provision of data used to be covered by national policies, national ownership and national copyright. This situation has improved with the new Europe-wide CORINE licence agreement, at least for research use of the CORINE land cover data.

The conditions of ownership typically restrict access to certain uses. This has given rise at the JRC to some environmental databases that are directly relevant to other research projects at the JRC, but which are not accessible because they fall outside the access conditions. The restrictions in a licence to a single application or a specific purpose is a common problem in many projects.

In the case of applications in Russia, the experience from the Siberia II project is that the relevant Russian forest institutions claim ownership of all data on forestry even where the European Commission has funded the updating and digitising of maps corresponding to approximately 250,000 ha of land. In Russia, maps of a scale better than 1:200,000 are regarded by the state as confidential, and a combination of the Federal Security Service (FSB) and Roskartografia reviews and controls all cartography at such scales.

A similar situation exists in China, but there the threshold map scale is 1:1 million. The Chinese national mapping service has extensive and up to date maps available at large map scales, but users who are not authorised by Chinese government departments can have access only to 1:1 million scale maps. In India, the cut-off scale for open availability of maps is 1:250,000. In several African countries the maps are owned by the state and regarded as state secrets. For example, it is not possible to buy maps of Congo in that country, although the EOLES (Earth Observation Linking SMEs to face real time natural disaster management) project has been able to purchase maps of Congo from the Belgian national mapping agency. In Tunisia, there are national topographic maps at 1:50,000 scale in existence, but these are not available for open sale because they are regarded as being in the military domain.

As well as ownership of databases, it is important to recall the need for access to data for calibration and validation purposes, and this can give rise to difficulties originating in privacy. For example, oil slick monitoring using Synthetic Aperture Radar (SAR) data may often require validation data held by oil companies; these oil companies keep much of their relevant data private and so the opportunity for validation may be frustrated.

A further complication is the possibility of competition, real or perceived. The Marine Institute in Greece needs access to atmospheric forecasts as boundary conditions for their ocean models. The European Centre for Medium-Range Weather Forecasting (ECMWF) is the best available source of these forecasts, but only six-hourly products are available outside the European national meteorological services. The Marine Institute in Greece has asked for higher temporal resolution data from the Greek meteorological service, but there are some difficulties because of the possible development of competitive weather forecasts.

INTELLECTUAL PROPERTY RIGHTS

The meteorological and oceanographic communities have made good progress with intellectual property rights (IPR) to encourage data sharing. In the case of meteorology, the World Meteorological Congress, at its meeting in June 1995, adopted WMO Resolution 40, which provides for free and unrestricted access to near real time meteorological data (Harris 2002). The core policy of Resolution 40 notes:

> As a fundamental principle of the World Meteorological Organization (WMO), and in consonance with the expanding requirements for its scientific and technical expertise, WMO commits itself to broadening and enhancing the free and unrestricted international exchange of meteorological and related data and products.

The resolution then expands on the policy by giving guidelines to WMO Member States on the practice of the resolution, including:

> Members shall provide on a free and unrestricted basis essential data and products which are necessary for the provision of services in support of the protection of life and property and the wellbeing of all nations, particularly those basic data and products as … required to describe and forecast accurately weather and climate, and support WMO programmes.

A similar situation exists for oceanographic data with the Intergovernmental Oceanographic Commission (IOC), at least for data relating to the open ocean (IOC 2001). There are, however, some obstacles to sharing of data in the coastal seas, for example the North Sea. Some biochemical data is sensitive and not shared, such as data on eutrophication associated with pollution discharges or the effects of accidental nuclear discharges into the sea. Data streamlining and open access is less mature for data of the shelf seas than for the open ocean, and this is particularly relevant to the major European seas – the Mediterranean Sea, the Baltic Sea and the North Sea.

IPR differ with the legal framework of different countries and with the European Commission's legal context. A good case in point is the Natura 2000 database. While the European Commission does hold the data, the Natura 2000 database is not yet available to research projects for the following reasons:

- the dataset is not yet complete;

- the site selection in some countries has been changing, particularly for the Special Protection Areas;

- for some countries the site selection is confidential; and

- the dataset is not yet fully validated.

Access to the Natura 2000 data held by individual Member States is possible through the Member States themselves, although from the BIOPRESS project experience the IPR vary between European Member States: for example, the UK Natura 2000 data is available on a website and is accessible at no cost, while the Finnish Natura 2000 data is available only for a fee.

The European Soils Database is not yet available because the IPR have not been agreed by the soil organisations of the European Union Member States. The European Soils Database exists at the JRC but, because of the restrictive approach to licensing from one Member State, the whole dataset of Europe has not been made widely available.

STANDARDS AND METADATA

Within the subject of standards and metadata there are three key themes that are important:

(1) the technical file formats and data storage standards;

(2) the standards that apply to the quality of the data as representations of reality; and

(3) the standards that relate to the procedures for information creation.

There is no convergence on standard ways of holding and transferring environmental data in the GMES projects. While there is some trend toward making data available in proprietary formats, for example ArcView, ERDAS Imagine or ENVI, there is much evidence of projects creating their own ways of holding data. This can be as simple as files containing rows and columns of byte data, or it can be projects writing their own programmes. The Land Degradation Assessment in Mediterranean Europe (LADAMER) project has no standard for its own data files within the project, but uses an ASCII ReadMe file to describe the data and products. The Marine Environment and Security in the European Area (MERSEA) project uses the Network Common Data Format (NetCDF) because of its advantages for handling flows of data across the Internet. NetCDF is widely compatible with other projects and data systems, for example with the Global Ocean Data Assimilation Experiment (GODAE) and the Argo ocean buoy network of profiling floats. Argo is an international effort to implement an array of 3,000 autonomous (free-drifting) temperature/salinity floats as a major component of a global ocean observing system. Argo floats measure temperature, salinity, pressure and reference velocity, together with sea surface height from satellite altimetric data. NetCDF may well emerge as a standard for oceanographic applications, although its application for terrestrial applications cannot be readily foreseen.

In many sectors of environmental science there are examples of good practice in calibration and validation. Forest inventory data used in the Siberia I and Siberia II projects for Germany and Russia is excellent. In the stratospheric research community, standards have been established to make worldwide co-operation straightforward. Data used in operational meteorology and oceanography have good calibration and validation processes. Inter-calibration is a major challenge for methane monitoring. While the National Oceanic and Atmospheric Administration (NOAA) in the US, and the Commonwealth Scientific and Industrial Research Organisation (CSIRO) in Australia have developed standards for laboratory greenhouse gas measurements, in Europe there is extensive use of inter-calibration among expert laboratories: without such inter-calibration the methane measurement data is of low value. There have been problems with European national data records: for example, the German national standards for methane emissions from landfill sites were incorrect by a factor of four, and in Belgium there are different calibration standards based on the region of Belgium in which the measurements are taken and analysed.

The procedures for generating environmental information and knowledge from data have not yet reached maturity in all sectors. VTT of Finland notes that the definition of a forest is not uniform in all countries. Under the Kyoto Protocol, each country has to produce an operational definition of forests, but these definitions can be substantially different. The procedures used by Norway and the UK to identify oil slicks in satellite Earth observation image data of the North Sea are different. Where aircraft data is used to monitor oil slicks, the characteristics of the aircraft and its sensor can have a substantial impact on oil slick detectability: different aircraft and sensors are used in Germany and France, for example. If aircraft perform other tasks, such as a coastguard function, the detailed information on location and other supporting data may be confidential.

In terrestrial applications, the use of proprietary Geographic Information Systems is increasingly common, but the data used in these systems may not be geometrically compatible. For example, the EOLES project found that the superposition of vector data of coastlines and basins in a Geographic Information System (GIS) showed that the data from different sources were not coincident. The problem is often increased when Earth observation data is included in a GIS.

Some meteorological datasets suffer from problems of discontinuity. The Siberia II project reports that in Russia many meteorological stations are closing, which means a poorer quality meteorological dataset because of the loss of the time series. Meteorological and hydrological data produced by Russia used to be a rich source of environmental information, but that richness is now reducing. Instead of Russian sources, the Siberia I and Siberia II projects have used meteorological data for the Siberia region from sources in the UK and in the US, namely the University of East Anglia and the University Corporation for Atmospheric Research in Boulder, Colorado. The data used by the European Seafloor Observatories Network (ESONET) project to create an atlas of the European margin is fragmented between different institutions, and does not necessarily adhere to any one standard: data older than just two years is often incompatible with updated software, or it requires specialist knowledge or equipment to produce meaningful information.

In the Earth observation sector there is some convergence of standards, at least for terrestrial applications. Common standards in widespread use include Geocoded Tagged-Image File Format (GeoTIFF) and Joint Photographic Experts Group (JPEG) image data formats, and the work of the International Standards Organization (ISO) and the Open GIS Consortium (OGC, recently renamed the Open Geospatial Consortium) offers a way forward for integration of data from a variety of sources. There can, however, be small technical problems that grow into major obstacles. The EUROSION project, for example, signed a memorandum of understanding with the Spanish spatial data infrastructure organisation IDEC. Both IDEC and EUROSION used ISO 19115 as the standard for indexing metadata. However, three small differences exist in the choice of mandatory and optional metadata fields, and these differences turned into major technical constraints to prevent IDEC and EUROSION data being easily interoperable.

Catalogue systems for describing data holdings have improved considerably in the last 10 years. An example of good practice is the ESA ODISSEO catalogue system (ESA 2004b). Because environmental research and applications projects require data from a variety of sources, there is great merit in providing front-end catalogue systems to a variety of data catalogues to provide easy user access to environmental data. The Information on Earth Observation (INFEO) system (INFEO 2003) was quoted by the EOLES project as a positive step to provide such multiple catalogue access.

LICENSING, DISTRIBUTION AND DISSEMINATION

The conditions attached to the distribution of ESA Earth observation data depend on the use of the data. The following two categories of use have been defined by ESA (1998):

> Category 1 use. Research and applications development use in support of the mission objectives, including research on long-term issues of Earth system science, research and development in preparation for future operational use, certification of receiving stations as part of the ESA functions, and ESA internal use.

> Category 2 use. All other uses which do not fall into category 1 use, including operational and commercial use.

In preparing this division of uses, the ESA Member States foresaw a clear distinction between pre-operational use of ESA Earth observation data and operational use of the data. In practice, the transition from pre-operational use (or preparation for future operational use) and operational use is not as clear as was at first thought. Some operational systems have a mix of fully operational elements and elements that are in the research and development phase.

The quality and availability of licences for access to data for research purposes are improving. For both CORINE land cover and CORINE biotope data access, the licence agreement for research use is openly available. In the case of the Statistical Office of the European Union (Eurostat), access to Geographical Information Service of the European Commission (GISCO) data presents some problems. Access is only free of charge if carrying out a project for Eurostat or for projects improving the dataset, so therefore the BIOPRESS project (which is an EC project but is neither a Eurostat project nor improves the GISCO data) was charged a fee for access to the GISCO data. In addition, access to the GISCO data is only available for internal project use and the data must be destroyed once it has been used in an approved project.

The licences for standard data products are typically under the control of the data supplier. When derived products are produced, there is some uncertainty over the extent of the coverage of the original licence, based on the characteristics of the derived product. Each Earth observation data provider has its own rules over the extent of the licence coverage, and becoming familiar with these different rules is time-consuming for project members, in particular for research groups and for Small and Medium sized Enterprises (SMEs). An interesting example is the Australian Centre for Remote Sensing (ACRES) which has a standard licence agreement (ACRES 2003) to cover data from ERS, JERS, RESURS and SPOT satellites. ACRES defines enhanced data, derived products and value-added products in its licence agreement as:

> ... data that has been processed to modify the information so that it can not be restored to its original form. The modification of the data can either be through the addition of information or through manipulation to significantly change the data. This does not include simple geometric and radiometric corrections.

A key question for uncertain users here is what is meant by the concept of significant change?

Data from the SRTM is a valuable source of topographic information for the land surfaces of the globe between 55°N and 55°S. The SRTM data has the following accuracy characteristics:

- Absolute horizontal accuracy : ±20 m

- Relative horizontal accuracy : ±15 m

- Absolute vertical accuracy : ±16 m

- Relative vertical accuracy : ±6 m

For users in Russia, these characteristics make the SRTM data fall, in principle, within the limit of confidential map data controlled by Roskartografia and the FSB. Potential Russian users, including Russian Principal Investigators and partners in European Commission projects who have the right to SRTM data at no cost, are deterred from placing orders for SRTM products of Russia as the data is perceived as falling within the guidelines of state confidentiality.

PRICING POLICY

Earth observation data is widely regarded as being too expensive. From a *user point of view*, data is expensive because its costs often either make up a substantial proportion of a project's budget or would make up a substantial proportion if the data was not provided for free. The Siberia I and II projects, for example, have been provided with data free of charge only because of the projects' participation in Announcement of Opportunity programmes that provide data at no cost. However, from a *supplier point of view*, the price of data is typically set at the level needed to capture only the marginal cost of data provision for government agencies or, in the case of commercial providers such as Space Imaging, to capture the real cost of data collection. There is no Earth observation data provider making a profit from data sales to the civilian sector, thus the perception of whether Earth observation data is expensive or not depends on the position of the viewer.

It is becoming increasingly clear that data provided free of charge often carries higher effort costs to the user than data supplied at an agreed price. The

EUROSION project has been trying to use the LaCoast dataset, which provides information on land cover change in European coastal zones between 1975 and 1990 at a map scale of 1:100,000. The LaCoast data is, in principle, accessible free of charge from the JRC but, after 18 months of trying, the EUROSION project failed to acquire this free data because there are no staff and computer infrastructure resources at the JRC to extract the data from the server, perform quality control operations, produce a CD and create a licence agreement for the data. Paradoxically, the EUROSION project did gain access to an old, unvalidated version of the LaCoast dataset that was openly available from the European Environment Agency (EEA). The ESONET project shares the view that, while some data carries no cost, there are substantial effort costs of free data and often no guarantee of the quality of the data provided anyway. The LADAMER project notes that while some datasets are available for free, there is a requirement for extra manpower to perform the electronic copying from the archive to other magnetic media.

The situation with non-Earth observation data is very varied, both by data type and by country. While aerial photography can be expensive, there is the benefit of owning the data, whereas for high spatial resolution satellite data the high cost is only for a licence to use the data rather than having full ownership of it. However, high spatial resolution satellite Earth observation data is available for most of the land surface of the globe, while the acquisition of aerial photography is subject to gaining over-flight rights of a territory.

In order to develop and test fully the procedures and techniques that could be useful for operational environmental monitoring, projects do need to use more data than just single images. For example, the MERSEA project needed daily SAR data of the North Sea, which at a cost of (say) €100 per scene means a total annual spend of €36,500. While for a robust operational programme this annual figure is not necessarily large, it is a significant cost for the MERSEA research project that is developing the features of an operational system. In addition, this should be set in the context of most *in situ* oceanographic data being made available for free. In the case of mapping wetlands, the extension from a small number of test sites, as in the European Commission Manhuma project (Manhuma 2000) and the TESEO project on the Ramsar Convention, to include all 1,400 Ramsar wetland sites around the world would represent a significant cost. The pricing policy issues of operational systems are challenges for the transition from the support of research to the support of sustainable operational services.

ARCHIVING POLICY

Some archives of environmental data are in good condition and thought of highly in the environmental community, for example Landsat data, CORINE land cover data and some aerial photography. The Network for Detection of Stratospheric Change (NDSC) maintains stratospheric ozone data from all over the world and ensures that data providers follow the same baselines for standards, metadata and methodology. Other archives suffer from a lack of a clear policy or from data discontinuity.

Access to data archives is vitally important for building models. For example, in the case of modelling atmospheric methane concentration, there is a need to have access to data on wind fields, atmospheric chemistry (in particular carbon monoxide and tropospheric ozone), wetlands, forests, biomass burning and pipelines. One lesson from environmental science of the last 20 years is the need to be able to reprocess data when models are improved or new analysis models generated. For example, because of the new gravity field data from the Gravity Field and Steady-State Ocean Circulation Mission (GOCE) and from the Gravity Recovery and Climate Experiment (GRACE), all radar altimeter data will benefit from reprocessing to produce a robust 20-year product of mean surface ocean height. It is likely that in the future it will be necessary to reprocess datasets approximately every 10–20 years in the light of new scientific advances.

There is great merit in designing archiving requirements into a project right at the start. Design for long-term archiving at the start of a project has a low cost; making modifications afterwards can be very expensive in cost and time. The EUROSION project agreed with the European Commission DG Environment, Eurostat and the EEA that, after project completion, the EUROSION data will be hosted within the GISCO database at Eurostat. Knowing beforehand that the data will be hosted at Eurostat made it possible to adapt the EUROSION data model to GISCO requirements, which helps to ensure a high level of compatibility, a wider community of users and a lower cost.

The challenge of archives will become more acute in the future. There will be more data, more models, more instrument modes, more types of media and more standards. While decentralised data archives have many attractions, there can be the problem of variable metadata. There is merit therefore in thinking about a central archive of environmental data, or at least a firm central control even if the physical locations of archives are distributed.

Reliable long-term archives are essential for climate change analysis, and for practical application projects such as land subsidence and land cover change analysis. The major concern for historical archives is the benefit of archives versus their costs. Benefits are often not measurable in financial terms. However, one might need an estimate of value to compare the archive benefits with the costs needed to run such an archive. Schreier (2002) has suggested how an Archive to Benefit Cost Ratio (ABCR) could be calculated for environmental data archives. The ABCR is the ratio between the total archive benefit, AB, and the total archive costs, AC. The higher the ABCR, the more it justifies the long-term maintenance of the archive. ABCRs below 1 might make a justification of the maintenance questionable. The ABCR increases with the time that unique data are kept compared to the real archive life; it is arguable that all environmental data is unique because the environment always changes, so measurements at a particular time are unique because the next measurements of the same phenomenon at the same location will be at a different time. The ABCR also increases with a greater investment in science, a greater profit from data sales and a greater proportion of the overall mission costs dedicated to post-mission archives.

CHAPTER 5

EVIDENCE FROM APPLICATION PROJECTS

INTRODUCTION

Suppliers of data publish their formal guidelines and policies to distribute data, but it is sometimes more difficult to obtain datasets in practice. Hidden licensing issues, incompatible data formats and other unexpected problems are only raised when a data user actually attempts to obtain the information. Consequently, the ease with which data can be obtained was investigated and is reported in this chapter, using versions of projects that might be implemented in global environmental monitoring systems. The types of data that may be required for such projects were identified and an attempt was made to locate and obtain the required datasets. The restrictions and data policy issues that were raised in the course of trying to obtain the data were then recorded. These examples are simpler than such projects would be in reality, but are useful in highlighting the ease with which different data types can be located and obtained in practice. The application projects are concerned with oil spills at sea, earthquakes and climate change. In addition, this chapter also examines the case of Natura 2000 data, as it has widespread application in Europe.

MONITORING MARINE OIL SPILLS

Objectives and data

Pollution associated with oil spills at sea can pose a serious threat to the environment as well as to local commercial interests. Some coastal areas are particularly vulnerable to the effects of oil pollution, coral reefs and wetlands being notable examples.

There are currently a number of oil slick observation initiatives in Europe and the rest of the world involving the use of radar or aircraft data for oil spill detection. The development of such monitoring and early warning capabilities for marine oil spills can help mitigate the effects of such events, and facilitate clean-up operations.

The objectives for any oil spill detection programme are likely to be as follows:

- to detect oil spills over a wide area in real time;

- to predict the likely behaviour and trajectory of an identified oil spill;

- to identify coastal areas, vulnerable ecosystems and habitats, and centres of population that could be at risk from an oil spill; and

- to identify potential polluters, for example ships, oil platforms or factories situated near the coastline.

The datasets investigated for monitoring oil spills are listed in Table 5.1.

Data type	Source	Comment
Satellite SAR data	Eurimage (European Remote Sensing Satellite (ERS) data), Radar Solutions (Radarsat data)	Can be used to detect marine oil slicks as the viscosity of the oil dampens the effects of waves and therefore returns a reduced signal to the sensor (see Figure 5.1)
Sea surface currents	National Oceanic and Atmospheric Administration (NOAA)	Needed for forecasting the trajectory of an oil spill
Wind fields	European Centre for Medium-Range Weather Forecasts (ECMWF)	
Wave height	ECMWF	
Oil type	*In situ* sampling	
Coastline locations	Mapping agencies, hydrographic agencies	Needed to assess the potential impacts of an oil spill on the local environment
Locations of environmentally sensitive areas and habitats	European Environment Agencies (EEA), national environmental agencies	
Locations of centres of population	EEA, national statistical institutes	
Knowledge of shipping lanes and routes	Hydrographic and maritime organisations	Identification of potential polluters
Knowledge of oil platform locations	Hydrographic and maritime organisations	
Knowledge of factory locations along the coast	EEA, national environment agencies	

Table 5.1: Data types for a project that monitors oil spills at sea

Data policy issues encountered

Satellite radar data

Radarsat data is owned and distributed by Radarsat International, and through Radar Solutions in the UK. ERS and Envisat data (see Figure 5.1) is available either through ESA directly for Category 1 use or through distributing entities for Category 2 use. The data cannot be downloaded or ordered online, but order forms can be downloaded. Whilst ERS data is free for research purposes, there are tight restrictions imposed on the use of such data. Additionally, acquiring data free of charge is subject to panel review that may mean that data cannot be acquired for several months after it is requested.

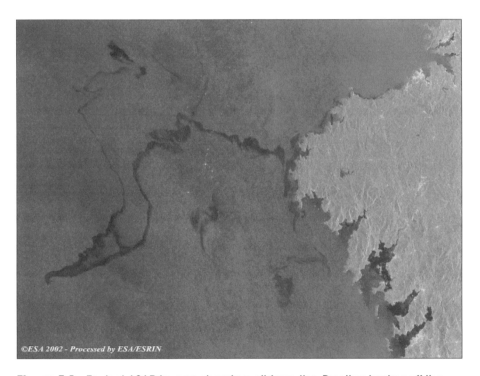

©ESA 2002 - Processed by ESA/ESRIN

Figure 5.1: Envisat ASAR image showing oil from the *Prestige* tanker off the coast of Spain. The black streaks are caused by the dampening effects of the oil spill on the sea waves. Source: http://earth.esa.int

Sea surface currents

One method for analysing sea surface currents is based on Ekman and geostrophic dynamics, using satellite-based, altimeter-derived sea level data and scatterometer-derived wind data. One source of sea surface current data is the Ocean Surface Current Analyses – Real time (OSCAR) project that is operated by the US National Oceanic and Atmospheric Administration (NOAA). This project is currently developing a processing system to provide operational ocean surface velocity fields from satellite altimeter and wind vector data. This data can be freely accessed from the OSCAR website (OSCAR 2004) in the form of a map, a time series plot, a latitude-time section or a time-longitude section. However, the project has a regional focus on the tropical Pacific, and would therefore be of no use for oil spills that occur in European waters.

Wind and wave data

Wind and wave data are produced by the European Centre for Medium-Range Weather Forecasting (ECMWF) and can be obtained from the national meteorological offices of Member States. In the UK, ECMWF data can be obtained through the British Atmospheric Data Centre's (BADC) website. Wind data can be acquired through the BADC website, although access is restricted so that it is only used for *bona fide* research purposes, and must be applied for in advance. Near-real time ECMWF data is restricted further, and access is granted on a case-by-case basis.

QuikSCAT real time Ocean Surface Wind Data can also be obtained on the Internet free of charge via anonymous File Transfer Protocol (FTP) (see Figure 5.2). QuikSCAT data is only available in Hierarchical Data Format (HDF), which may be a problem if the user does not have the required reading software.

ECMWF data is free of charge for research purposes under World Meteorological Organization (WMO) Resolution 40, subject to conditions, although marginal costs have to be paid. QuikSCAT data is free and easily accessible to download via anonymous FTP.

Location of coastlines

One source of coastline location data is the World Vector Shoreline data produced by NOAA. The accuracy of this product is said to be that 90% of all identifiable shorelines are located within a 500 m circular error of their true geographical positions, based on the World Geodetic System 1984 (WGS84). The data is available on a CD-ROM that can be ordered for US$75, or downloaded for free via the website.

Similarly, the Geographic Information Service of the European Commission (GISCO) Seamless Administrative Boundaries of Europe (SABE) coastline dataset can be used to delimit coastline locations. The data is available from Eurostat in vector format, and is supplied at a map scale of 1:100,000. The data has copyright protection and users must agree not to make copies of the data or make it available to third parties.

Locations of environmentally sensitive areas

Areas that may be particularly vulnerable to an oil slick can be identified using the Common Database of Designated Areas, which contains the geographic location and size of any nationally designated areas. The European Topic Centre on Nature Protection and Biodiversity maintains this database and Dublin Core metadata to

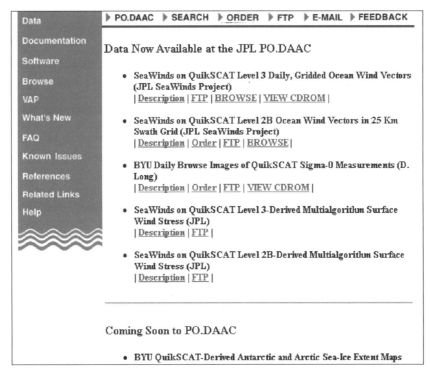

Figure 5.2: Online data portal for QuikSCAT data. Source:
http://podaac.jpl.nasa.gov/quikscat/qscat_data.html

describe the data is available on the website. The data is free, accessible through the Internet and is downloadable in a variety of formats such as ARCInfo, Access and Dbase.

The Co-ordination of Information on the Environment (CORINE) Biotopes dataset can also be used to identify areas of environmental importance that may be vulnerable to an oil slick. This dataset is an inventory of major nature sites in Europe and provides information on vulnerable ecosystems, habitats and species. The data can be downloaded free of charge from the European Environment Agency (EEA) Data Service website free of charge, and is supplied with Dublin Core metadata. The user must submit a signed user agreement in order to download the data.

Location of centres of population

Population density and land cover in coastal areas in Europe can also be easily downloaded from the EEA Data Service, and forms part of the CORINE coastal erosion database. Coastal population data is also free and easily accessible through the EEA Data Service in the form of a detailed map (see Figure 5.3). The coastal erosion database can also be downloaded.

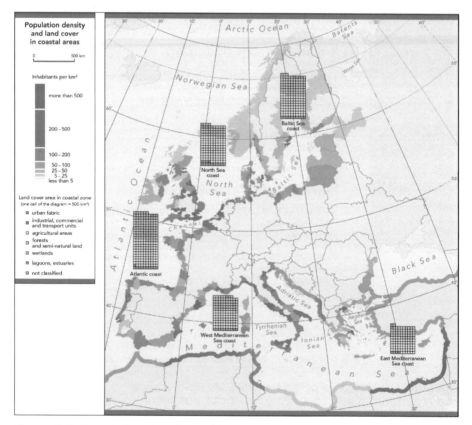

Figure 5.3: Population density and land cover map for coastal areas.
Source: EEA Data Service http://dataservice.eea.eu.int/dataservice

Knowledge of shipping lanes and oil platform locations

A suitable source of information about shipping lanes and oil platforms is the United Kingdom Hydrographic Office. Nautical charts and information are available in digital format (on CD-ROMs) and hard-copy format. Charts and digital products can be obtained through a network of authorised distributors at the market price, although data is Crown copyright and cannot be copied or reproduced without permission.

Knowledge of factory locations along the coast

Information describing land use in coastal regions in Europe, including industrial and commercial units, is available as part of the CORINE land cover dataset. The information is available as either a raster 250 x 250 m database, or as a 100 m resolution vector dataset. The data is freely available from the EEA Data Service, but requires the user to sign an agreement form before the data can be downloaded. This agreement requires that the data is not used for commercial purposes or distributed to third parties.

Conclusions

It is not always easy to discover which information is available for oil spill monitoring and under what conditions it can be acquired and used. The data types described above came from a variety of sources, many of which required extensive searching to uncover. More suitable datasets than those described above may exist for oil spill monitoring, but may not be in the public domain or may not have adequate descriptions and metadata to check their suitability for the purpose.

Information is not always abundant and many datasets have conditions that constrain their extensive use. Many of the datasets also had conditions governing their use and acquisition, which may prove restrictive for an oil spill monitoring project. Radar data is relatively simple to acquire; the main issue that may limit access is cost. In addition, there is no online ordering or downloading system available for Radarsat or ERS data. In an application such as oil spill monitoring, which would require data in near-real time, a lack of immediately available data could be a serious obstacle. Additionally, the conditions by which ERS data can be obtained free of charge for research purposes are tightly controlled and complicated. If research access is not granted, the data has to be paid for, which might hinder the progress of some applications. The EU Forum on Earth Observation Use for Environment and Security (EUFOREO) report (Cannizzaro 2003) notes that an oil spill monitoring service would require large volumes of input Synthetic Aperture Radar (SAR) data but the current pricing policies of suppliers do not offer sufficient price attraction for volume orders. The EUFOREO report also notes that the revisit time of current single operational satellites is too low and cannot provide daily coverage of all areas of interest for an oil spill monitoring scheme. Frequency of coverage has been improved recently however, with the availability of Envisat data.

Nautical charts can be obtained in the form of paper maps and digital data from agencies such as the Hydrographic Office of the United Kingdom. Other issues, such as intellectual property rights (IPR) in the case of UK Hydrographic Office data can limit, or at least complicate, the freedom and the degree to which information can be used.

Additional supporting information, such as that used to forecast a spill trajectory, comes from a variety of sources, and its accessibility is variable. For example, Ocean Surface Wind Data from the QuikSCAT satellite is freely available for anyone to download via FTP. By comparison, wind data derived from ECMWF data is free, but access is restricted so that it can only be used for *bona fide* research, and its use can be monitored.

The accuracy and scale of some spatial datasets may be unsuitable for applications such as oil spill monitoring. The mapping and topography data needed for such a project was relatively easy to acquire through NOAA. World Vector Shoreline data was easily obtainable on CD-ROM for a reasonably low cost of US$75, or it can be downloaded through the NOAA website. Accuracy and scale are likely to be the main issues regarding this data for oil spill monitoring. World Vector Shoreline data is said to be accurate to within a 500 m circle of their true location, and suitable for map scales of close to 1:250,000. For oil spill monitoring on a local level at 1:25,000 map scale, this may not meet the required accuracy.

Some datasets are already available for access and viewing free of charge, are accompanied by descriptive metadata and are easy to locate. The NASA QuikSCAT data was available free of charge and easily downloaded from the NASA Jet Propulsion Laboratory website (via FTP). In addition, the data was described in detail and the user can browse all quick-look images of the data before downloading.

The EEA Data Service also offers easy access to a wide variety of data, and relevant datasets showing designated areas and coastal population and land use were easily obtained and downloaded. There were also detailed descriptions accompanying this data, and it was available in a format that was easy to understand and interpret.

EARTHQUAKE MONITORING AND MITIGATION

Objectives and data

Earthquakes can be a particularly devastating natural phenomenon, and often result in severe social and economic consequences. Earthquake monitoring programmes could be a valuable tool in mitigating the effects of such events.

The main objectives of an earthquake monitoring programme might be as follows:

- monitoring seismic activity and crustal deformation;

- identification of areas most at risk from the effects of earthquakes; and

- provision of information and support for emergency services and government after a disaster occurs.

Table 5.2 summarises the types of data that might be required to meet these objectives.

Data policy issues encountered

Satellite radar data

SAR data for interferometry requires that two images from different times are used, but the satellite track and look angle must be as similar as possible. Users are able to request baseline information from Radarsat International for pairs of scenes of interest, so that the suitability of the data can be checked before ordering. Radarsat data costs between US$100 to US$1,000 per scene, depending on the delivery time, whilst ERS data costs around €1,000–1,400 depending on the level of processing.

Data type	Source	Comment
Satellite SAR data	Eurimage (ERS data), Radar Solutions (Radarsat data)	SAR interferometry to measure the movement of land features
Satellite thermal infrared data	NASA Land Processes Distributed Active Archive Center (DAAC) (MODIS and ASTER data)	Thermal anomalies on the ground can be detected before an earthquake event, although the technique is not fully developed
Seismic station data	European Mediterranean Seismological Centre, International Seismological Centre, seismological centres of individual countries	Calculation of location, magnitude of earthquakes
Geological and tectonic maps and data	National Geological Survey institutes, mapping agencies	Identification of seismic areas
Maps and plans of buildings and infrastructure	Utility companies, local authorities and planning authorities	Aiding and directing emergency operations after an earthquake, identifying areas at risk from gas leaks
Population data, maps of population distribution	National statistical organisations, census data, local authorities	Identifying areas likely to contain the most casualties after an earthquake, supporting emergency services

Table 5.2: Data types for an earthquake monitoring project

Thermal infrared satellite data

Thermal infrared data from sensors such as the Moderate Resolution Imaging Spectroradiometer (MODIS) is reasonably easy to obtain. In the case of MODIS, data is free and there appear to be little or no restrictions on its use. MODIS data can be ordered over the Internet from the Land Processes Distributed Active Archive Center (DAAC). Becoming a registered user is a simple process, and only requires the completion of a simple online application form. Data can then be ordered, and an email is sent when it is ready to be downloaded by FTP or similar method. Advanced Spaceborne Thermal Emission and Reflection Radiometer (ASTER) data can also be ordered through this method, although there is a charge for data. Whilst MODIS data is free, ASTER data costs US$55 per granule, plus a US$5 handling fee.

Seismic station data

Seismological data can be obtained online from the European-Mediterranean Seismological Centre (EMSC) and the International Seismological Centre (ISC). Real time seismicity data is available from the EMSC website (EMSC 2004), as is access to several databases. This includes the Strong Motion database, the Euro-Mediterranean Intensity Database, and Instrumental Seismological Databases. Data is available on the EMSC and ISC websites with no restrictions on access. Data can be downloaded in a variety of formats from EMSC and is also displayed on the website, and ISC data can be accessed as a searchable online database. However, real time seismicity data from EMSC consists of automatic detections conducted by European networks, and their use for scientific research purposes is strongly discouraged due to issues over accuracy.

Geological and tectonic maps

Geological maps can be obtained from national geological agencies such as the British Geological Survey (BGS). IPR to this data are held by the Natural Environment Research Council (NERC) and data cannot be reproduced without permission. Discovery metadata with a description of all of the BGS's products is available online, and products can also be ordered from the Internet at market prices. The United States Geological Survey (USGS) has an online database of geological maps containing a description and where to find the data, and some maps can be downloaded directly from the USGS website via the US National Geologic Map Database.

Maps of buildings and infrastructure

A source of detailed infrastructure maps and information could not be found in the public domain. Such documents are held by bodies such as gas and electricity companies and city planning authorities, but are not made routinely available to the public. Very high resolution satellite data, such as that from the Ikonos and Quickbird platforms, could be used to provide accurate information on buildings and land use in order to identify those areas where casualties are likely to be high. However, the high cost of this data may be prohibitive and the data is not sufficient for locating gas mains, electricity pylons or water and sewer systems.

Population data

Information on population can be acquired via national statistics agencies, many of which have comprehensive online services from which data can be obtained free of charge (see Chapter 3). Much of the data available from statistics agencies is at regional levels and fairly general. Data that gives the population distribution in specific neighbourhoods or streets, for example, is not widely available. Such information is more likely to be held by local councils, although this could not be easily accessed or found online. National statistics agencies often provide more detailed, tailor-made statistics to order, which are charged at a price reflecting the number of man-hours taken to produce the data and the material costs. In some countries, such as the US, population information can be acquired down to the street level through the census bureau. However, this level of detailed census information is not freely available in the majority of European countries, and individual privacy is an issue for census data that is available for small areas.

Conclusions

Many spatial datasets relevant to earthquake monitoring have restrictions that hinder the sharing of data between users and applications. For example, map and hydrological data are often tightly controlled in order to protect IPR. Reproduction of data and the use to which the data is put is restricted and subject to licence. By contrast, national statistics data is disseminated freely and can be reproduced as long as the source is cited.

Spatial data and information such as geological maps also vary in terms of price and accessibility, depending on their source. For example, the BGS provides geological maps in paper and digital format that can be ordered online for the market price. By comparison, the USGS provides an online geological map database that contains descriptions of a large number of geological maps covering the entire US region. Many of these maps are available online and can be downloaded free of charge.

Information is often only available at a general level. National statistical institutes provide a large amount of free data via the Internet, although this information is generally only available at regional, not local, levels. Most datasets that are currently freely available tend to be at a more general level, whilst information needed for detailed investigations is much harder to discover and acquire. Information suitable for earthquake monitoring, detailing the number of occupants in individual buildings, for example, is difficult to obtain. Census bureaus can be useful information sources: for example, the US Census Bureau provides a data portal that allows a user to search and obtain data down to street level. Local authorities could be a source for such data, although they are less likely to have the resources to disseminate such statistics via the Internet.

The required geographic information was not always easy to discover. Much of the information required for an earthquake monitoring project was unearthed only after extensive searching, and some could not be located at all. In addition, the degree to which datasets were discovered and available was often dependent on the country or region to which the data referred. For example, US data and information is often easy to discover and is made available online. Geological data for European countries is not available from a central source, but is often accessed through mapping agencies of the individual countries.

Earth observation data is easy to find and obtain via the Internet, although its accessibility and cost is variable. For example, some data provided by NASA (for example, MODIS) is free of charge and can be downloaded directly from their servers. Radarsat and ERS data has much more restricted access, and has to be paid for if it is to be used for commercial purposes.

For many datasets, the standards and data quality procedures to which it was produced was not available. For example, real time seismic data obtained from EMSC was produced by automatic networks and its use for scientific investigations was discouraged. In addition, many of the datasets suffered from a lack of adequate metadata to assess their suitability.

CLIMATE CHANGE

Objectives and data

The environmental, social and economic consequences of climate change are likely to be significant. There has consequently been a large amount of research in climate change involving international organisations such as the World Climate Research Programme (WCRP), the International Geosphere Biosphere Programme (IGBP) and the Intergovernmental Panel on Climate Change (IPCC). The large amount of research, combined with the fact that climate change is a complex process involving many variables, has meant that there is consequently a large amount of data with varying levels of accessibility.

The basic objectives of any investigation into climate change are likely to consist of establishing which parameters are responsible for climate change and then measuring these parameters to attempt to model and predict future changes. Whilst this case study does not represent the complexity and depth of all climate change analyses, it is useful in establishing the data policies of data suppliers that may be associated with climatic data.

Kellogg and Bojkov (1983, cited in Kondratyev and Cracknell 1998) suggested the following order of priority for data on factors of interannual climate variability: (1) atmosphere-ocean interaction; (2) deforestation; (3) variability of snow and ice cover extent; and (4) other factors including urbanisation, carbon dioxide, tropospheric aerosols, desertification, stratospheric aerosols and soil moisture. On a timescale of decades, the priority might be as follows: (1) carbon dioxide; (2) deforestation; (3) urbanisation, atmosphere-ocean interaction; and (4) other factors including aerosols, solar activity, volcanic eruptions, stratospheric ozone, anthropogenic heat releases and snow and ice cover. The observations and data needed for researching climate change are listed in Table 5.3. The amount of funding and research that is attributed to climate research meant that the majority of datasets were discovered easily, and in many cases were produced by large, international organisations.

Data type	Environmental variable	Source
Basic meteorological parameters	Temperature Wind speed Relative humidity Surface pressure Sea surface temperature	ECMWF, National Meteorological Services
Radiation budget and components, cloudiness	Solar constant Extra atmospheric ultraviolet solar radiation Earth's radiation budget Cloudiness Global radiation Net longwave radiation Surface albedo	Goddard DAAC, BADC
Oceanic parameters	Sea surface temperature Upper-layer heat content Wind stress Ocean surface level Surface currents Deep water circulation	Global Ocean Observing System (GOOS), British Oceanographic Data Centre (BODC)
Precipitation and hydrology	Precipitation over ocean River runoff Soil moisture	Global Precipitation Climatology Project (GPCP)
Cryosphere	Snow cover extent Sea ice extent Ice thickness Sea ice melting Sea ice drifting Thickness of glaciers Deformation of glaciers Variations in glacier boundaries	Global Terrestrial Observing System (GTOS)/Global Climate Observing System (GCOS), Radarsat, SMMR and Special Sensor Microwave/ Imager (SSM/I) data, National Snow and Ice Data Center (NSIDC)
Atmospheric composition	Carbon dioxide Vertical profile of ocean concentration Total ozone content Global distribution of ozone Tropospheric aerosols Atmospheric turbidity Stratospheric aerosols	BADC, GCOS
Terrestrial parameters	Fraction of Photosynthetically Active Radiation (FPAR) Leaf Area Index (LAI) Land cover/land use	NASA land processes data archive, Joint Research Centre (JRC)

Table 5.3: Data types for climate change research (adapted from Kondratyev and Cracknell 1998)

Data policy issues encountered

Projects and data

Data for climate change analyses can come from or through the Global Ocean Observing System (GOOS), the Global Precipitation Climatology Project (GPCP), the ECMWF and from national meteorological services. The nature of the topic means that many of the datasets are from Earth observation satellites, as well as *in situ* monitoring networks. The Goddard Distributed Active Archive Center (DAAC) produced a dataset known as the Climatology Interdisciplinary Data Collection that covers almost all of the parameters listed in Table 5.3, and was designed for the study of global change, seasonal to interannual climate change, and other phenomena that require many interacting parameters. The collection comprises datasets covering atmospheric dynamics and atmospheric sounding products, radiation and clouds, biosphere data, measured atmospheric constituents, measured surface temperature and pressure, hydrological data and remote sensing science products. This dataset is freely available for browsing purposes on the British Atmospheric Data Centre (BADC) website (BADC 2004), and the data itself can be ordered on CD-ROM from the Goddard DAAC. Many of the datasets are also freely available by anonymous FTP via the Goddard DAAC's website (DAAC 2004).

Meteorological parameters

Basic meteorological parameters such as wind speed, humidity and surface pressure can be obtained from ECMWF and national meteorological services. Access to this data in the UK is through the BADC. Data from the BADC can be accessed and downloaded from the website (see Figure 5.4), although for some datasets, including those from ECMWF, access is restricted to *bona fide* research purposes only. In order to access the datasets, the user must apply for access and follow certain access regulations:

- The BADC can only distribute data to academic users within the UK; other countries have to apply through their own national meteorological service.

- The dataset can be used for academic research only.

- Users are required to sign a NERC-Meteorological Office agreement form that is then sent to the BADC.

- Users must also sign a dataset specific agreement form and send this to the BADC.

Limited observational data can be obtained from the UK Meteorological Office free of charge through its website, although more detailed datasets have to be applied for. For higher education projects that do not receive any additional funding, commercial fees are waived as long as the results are released into the public domain. However, projects receiving funding from industry, commerce, government or the EU must anticipate paying for the data at standard rates. In order to establish in which category the user falls and to gain permission to

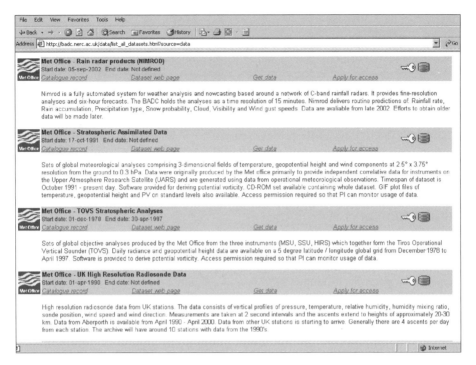

Figure 5.4: Part of the dataset index from the BADC website. Source: http://badc.nerc.ac.uk

acquire the data, the user must fill in an academic research application form and return it for a quotation.

Radiation budgets and components, cloudiness

Many of the datasets required for the Earth's radiation budget can also be acquired through the BADC. These datasets include the Earth Radiation Budget Experiment data and the Surface Radiation Budget data created by the World Climate Research Programme (WCRP). Data is also available from sources such as the Solar and Thermal Atmospheric Radiation (STAR) project of the NOAA Climate Monitoring and Diagnostics Laboratory, which specialises in the investigation of climatically significant variations in long-term radiation and meteorological measurements. Access to digital data is only through direct request to the STAR group, although all the data currently being collected is made available in a graphical form free of charge via the website.

Cloudmap2 data is also available from the BADC website, although it is currently only available for members of the Cloudmap2 consortium. It is envisaged that the data will become available to the wider scientific community in the next few years after thorough validation by the scientific team.

Oceanic parameters

Many of the oceanic parameters required for climate change research can be obtained through organisations such as the BODC and the GOOS. Several global datasets are also available through the US National Oceanographic Data Center. For example, global sea level information derived from satellite altimeter data can be freely downloaded by anonymous FTP, and satellite sea surface temperature and Argo buoy data is also available on the Internet. All datasets are provided free of charge, although customised data requests, CD-ROMs and publications may have an attached user fee. GOOS data is available from the Global Observing Systems Information Centre, and most of these datasets can also be downloaded free of charge.

Precipitation and hydrology

Global precipitation data can be collected using meteorological satellite data and rain radar. Global precipitation data estimated from Geostationary Operational Environmental Satellite (GOES), Geostationary Meteorological Satellite (GMS), Meteosat and NOAA satellites and combined with rain gauge data is available from the GPCP. Data from the GPCP is available from the World Data Centre (WDC) for meteorology via anonymous FTP, free of charge and without restriction.

A useful source of soil moisture data is the Global Soil Moisture Data Bank operated by the State University of New Jersey. Datasets contain *in situ* observations from a large number of sites throughout Asia, Russia and the US, although not Europe. Data is free to download via anonymous FTP, although the data providers ask to be notified if a user downloads a dataset.

Cryosphere

An excellent source of snow and ice data is the NSIDC, which houses many datasets covering cryospheric parameters. For example, there are several datasets covering sea ice extent, including MODIS/Aqua sea ice extent data, Defense Meteorological Satellite Program (DMSP) SSM/I daily polar sea ice concentrations data, and Arctic and southern ocean sea ice concentrations. The centre provides access to datasets covering ice growth/melt, sea ice motion, albedo (derived from AVHRR data) and ice thickness.

Most of the datasets can be obtained free of charge via anonymous FTP from the NSIDC website (NSIDC 2004). However, in some cases data is provided on CD-ROM (digital SAR data of the Greenland ice sheet, for example), or is accessed through another website by means of a hypertext link (for example, the Earth Observing System (EOS) Data Gateway for MODIS data). An online search tool is also provided on the website so that users can quickly locate datasets suitable for their purposes.

Atmospheric composition

As with many of the other parameters relating to meteorology and the atmosphere, datasets relating to atmospheric composition are available through national meteorological services. In the UK, datasets regarding aerosols, atmospheric turbidity and ozone can be obtained through the website of the BADC.

Terrestrial parameters

LAI and the FPAR data are available as a MODIS/Terra product that can be freely acquired from the EOS Data Gateway. An example of a LAI/FPAR product is shown in Figure 5.5. This product is derived from the atmospherically corrected surface reflectance product, the land cover product and ancillary information on surface characteristics using a 3D radiative transfer model. The data and the accompanying description, user guide and Algorithm Theoretical Basis Document can be ordered and downloaded free of charge from the EOS Data Gateway.

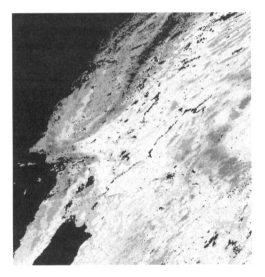

Figure 5.5: An example MODIS/Terra LAI/FPAR product. Source: Earth Observing System

Global LAI data can also be acquired as field measurements from the Oak Ridge National Laboratory DAAC, covering the period 1932–2000. This dataset is comprised of approximately 1,000 published estimates of LAI from nearly 400 field sites, together with associated variables such as latitude, longitude, elevation, stand age, and above-ground Net Primary Productivity (NPP) (Scurlock *et al* 2001). The data are available as a spreadsheet and an ASCII file, plus a bibliography of over 300 original source references and a number of companion files. A world map showing the distribution of field sites is also available. This data can be downloaded free of charge from the Oak Ridge National Laboratory.

Land cover data is available from the Global Land Cover Characteristics Database developed by the USGS, the University of Nebraska-Lincoln and the European Commission's Joint Research Centre. The dataset has been created at 1 km resolution and is based on Advanced Very High Resolution Radiometer (AVHRR) data. The dataset was developed, and can be downloaded, on a continent-by-continent basis. This data is also provided free of charge via the Land Processes DAAC, along with relevant documentation and information relating to the dataset.

Conclusions

The types of data that were identified for a climate change project were generally easy to locate and access, and in many cases were available through the Internet. For example, datasets specifically related to climate change are available through the BADC in the UK, and these can be accessed through the website of the organisation. It was similarly easy to locate datasets relating to the cryosphere, oceans and meteorological parameters, and many were produced by the same, often publicly-funded, agencies (for example, NOAA in the US).

Most of the datasets also had unrestrictive access conditions, and could often be downloaded free of charge. For example, many of the datasets located in the US, such as those produced by NOAA or derived from NASA satellites, were free to download via anonymous FTP. In cases where datasets could only be provided on CD-ROM or as a publication, marginal costs were charged. In the UK, most of the datasets were free but required the user to complete an application form before the data could be accessed. This policy was used to ensure the data is only used for *bona fide* research purposes and could not be exploited for commercial use.

Most of the data appeared to be of a high standard, produced using sophisticated techniques and rigorously validated. The majority of datasets were accompanied by detailed descriptions that described the quality, temporal and spatial resolution and a variety of other information relating to the data.

ACCESS TO NATURA 2000 DATA

Natura 2000 network

The Natura 2000 ecological network was initiated under the European Commission Directive (92/43/EEC) on habitats. When complete, it is likely to include between 12,000 and 15,000 sites, covering between 10% and 15% of the territory of the European Union (in 2000), and is designed to protect European biodiversity thanks to the coherent and comprehensive coverage of all habitats and species of interest. Natura 2000 sites are split into two categories: Special Protection Areas (SPAs) for the protection of wild birds, and Sites of Community Importance (SCIs) for the protection of wild flora and fauna.

SPAs are designated individually by Member States. SCIs are initially proposed by Member States and, after assessing and discussing proposals, a list is approved by the European Commission. Within six years of adoption of SCIs, each Member State must put in place necessary measures to protect and manage the sites. From then on they will be designated as Special Areas of Conservation (SAC).

Data about the sites is compiled by the responsible national authorities and submitted to the European Commission. This submission comprises the following information:

- a paper map and data form for every site;

- a descriptive database in MS Access format for information on sites; and

- a digital spatial dataset of the sites.

A Geographic Information System (GIS) project has been created to check and assemble the national contributions into an operational database and to develop a system to exploit data and make it available to European Commission officials and other users. The components of this system play two key roles: first, they provide a mechanism for harmonising and validating the incoming data from Member States. Secondly, they provide analytical tools to model, monitor, visualise and publish data relating to Natura 2000 sites (Vandenbroucke and Peedell 2003).

There have been numerous problems encountered in the creation of the Natura 2000 GIS and data system, particularly when receiving data from Member States. These problems include:

- spatial data that is often in different formats;

- a lack of meta-information;

- use of different map projection systems and a lack of information on the parameters used; and

- site delimitation defined on the basis of different kinds of topographic maps.

However, the most difficult problem is related to the validation and quality assessment of the data itself. Initial validation is required of paper maps and digital spatial data provided by a Member State. The first validation is needed since the paper map remains the only legal reference document, and the digital information system can only be used when the digital sites correspond exactly with the paper information.

Validation then comprises the following steps:

- verification of the completeness of maps to ensure that, for each site, a map has been sent to the European Commission;

- verification of the usability of paper maps for validation of spatial digital data (for example, well-known reference points, map projection information);

- verification of the completeness of the technical information for the spatial digital data (for example, map projection information, site coding).

Sites that are found to have differences between the different sources of information for the area, position and shape of the site are then reported as being potentially erroneous.

Natura 2000 data policy issues

The whole Natura 2000 dataset is currently not released due to reasons of confidentiality, and Member States have to agree on the dissemination of data on an individual basis. Although data can be obtained through individual Member States, the availability of Natura 2000 site information can vary considerably between countries. For example, the information for the UK is freely available on the Internet, including a downloadable version of the GIS boundary data. Finnish data can be purchased and has a licence that controls the redistribution of data, and in Spain it is virtually impossible to obtain coherent data and co-ordinated information about Natura 2000 sites.

The reluctance to distribute the Natura 2000 data can be partly explained by the fact that the implementation of the Natura 2000 network has been plagued by difficulties and delays. As a result, site selection for the network was completed only at the end of 2003, instead of in 1998 which was the date initially foreseen. Such problems can be traced back to the national level, particularly in some countries where site selection is deemed to be inadequate or controversial (WWF 2003). The delays can be explained by local concerns that Natura 2000 designation will damage prospects for economic development in these areas, with biodiversity protection being seen as implying costs or restrictions to local people.

Implications for GMES

The problems associated with obtaining Natura 2000 data have a number of implications for Global Monitoring for Environment and Security (GMES), particularly for GMES projects that require Natura 2000 data for their research. BIOPRESS, for example, has not been able to acquire the Natura 2000 dataset from the European Commission Directorate General (DG) of the Environment that holds the data, and has had to perform queries and analyses at the DG Environment consultancy office in Belgium.

The example of Natura 2000 highlights some clear data policy issues that affect GMES as, even though the European Commission produces the data, other projects funded by the Commission cannot get access to the data. This problem is at least partly due to policy issues between individual nation states, some of which are reluctant to divulge the exact location of the designated Natura 2000 sites. The case of Natura 2000 shows that there is a need for a strong collaboration between the European, national and regional levels to cope with data issues, especially over exchange of data, standards and procedures (Vandenbroucke and Peedell 2003).

CHAPTER 6

EVIDENCE OF GOOD PRACTICE

INTRODUCTION

Europe has not been at the forefront of environmental information systems at regional and global levels. There is extensive national expertise, especially at the research level, but Europe itself cannot claim to be a leader in spatial data infrastructures for dealing with environmental data from a wide variety of data sources. This chapter reviews the experience from Australia, New Zealand, Canada and the US in spatial data infrastructure, to draw lessons on the efficient and effective development of systems for global environmental monitoring.

AUSTRALIA AND NEW ZEALAND

ANZLIC

Australia is a leader in the field of national spatial data infrastructure development, in the form of the Australia and New Zealand Land Information Council (ANZLIC), an umbrella body that represents different levels of government. The broad approach taken by ANZLIC is to maximise the benefit to the community by providing better access to holdings of spatial data, at the marginal cost of transfer, in order to maximise the net economic and social benefits arising from use of the data. These benefits are maximised when (ANZLIC 1999):

> … the community has easy, efficient and equitable access to spatial data in an environment where technology requirements, data formats, institutional arrangements and contractual conditions do not inhibit use.

The terminology of maximising the net economic and social benefits is worthy but rather imprecise. The benefits have been made more precise by ANZLIC in recognising the main drivers at the national level for a spatial data infrastructure. These drivers are: actions needed to counter terrorism; the management of emergencies and hazards; marine monitoring and control, particularly over marine boundaries; and natural resources management. Spatial data infrastructure can then be regarded as an enabling infrastructure to respond to the requirements of these four drivers.

ANZLIC has adopted a number of strategies in order to achieve its aims, such as advocating the use of common standards, ensuring that data is more easily available to decision-makers and increasing the range of spatial information that is available to government, business and the community. In order to implement these strategies, ANZLIC is aiming for the development of a spatial data infrastructure with the following characteristics (ANZLIC 1999):

- a network of jurisdiction and agency-based databases that satisfy the community's need for nationally-consistent fundamental datasets;

- technical standards endorsed by ANZLIC and, where appropriate, submitted to Standards Australia for consideration as national standards, facilitating the sharing of data between agencies and jurisdictions and providing the necessary consistency and compatibility to enable fundamental datasets to be combined to develop value-added products;

- intergovernmental arrangements to facilitate the equitable sharing of data between agencies and jurisdictions;

- administrative arrangements and policies that facilitate industry and community access to data, under conditions that promote better decision making based on good quality fundamental spatial data and the development of a competitive spatial data industry; and

- an Australian Spatial Data Directory (ASDD), implemented as a distributed network of jurisdiction and agency-based directories, complying with standards endorsed by ANZLIC.

ANZLIC focuses on three enablers for its spatial data infrastructure: consolidation of capabilities; encouragement of a strong commercial sector; and encouraging people and good practice. The five priority themes in ANZLIC are: governance; access to data and services; data quality; interoperability; and integrate-ability. ANZLIC works by identifying areas of common interest and then seeking ways to improve capability and performance in each area. Along with ANZLIC, the states of Australia have set up a private company, Public Sector Mapping Agencies Australia Ltd (PSMA). Each state is a shareholder in PSMA and the company has a capacity for developing national mapping products. PSMA has agreements with the states in Australia and pays royalties to the states for data use.

The policy of ANZLIC forms the basis of the national data policies in Australia and New Zealand regarding the use and dissemination of spatial data. The implications of this policy, in terms of the six data policy characteristics used in the earlier chapters of this book, are described below.

Ownership, privacy and confidentiality

The Australian government views spatial data as key to sustainable management and planning at national, regional and local levels. A key concept of the Australian Spatial Data Infrastructure (ASDI) is 'custodianship', where responsibility for fundamental datasets is attributed to a particular organisation. A variety of organisations and bodies are therefore responsible for data collection and maintenance within the ASDI. Ownership of fundamental spatial datasets and associated property rights remains with the individuals or organisations that produced them, although free access is granted provided the interests of the owners are protected. There are two categories of access to spatial data: community access and restricted access. Community access generally applies to most government-funded projects, and data is made freely available to other parties and the public. Restricted access applies to data that may be protected by confidentiality and use provisions, and only made available to interested parties by agreement of the data owner on a case-by-case basis. If data is restricted because of confidentiality requirements, those responsible for the data may make it available for public access as an aggregated dataset. ANZLIC aims to reclassify restricted data over time, subject to approval by the data custodian or owner.

Intellectual property rights and associated legal frameworks

Intellectual property rights (IPR) remain with the producers of the dataset, even if the data is made freely available. Products derived from the data must contain an acknowledgment of the source, data quality statements and any disclaimers required by the custodians.

Standards and metadata

Datasets are developed and maintained to meet agreed international or national guidelines or standards as endorsed by ANZLIC, or through national co-ordination arrangements. This is to ensure that datasets are comparable and consistent where required, and can be used for a variety of applications. Standards being implemented by ANZLIC include structure and content for spatially referenced data, metadata guidelines, geocoding and addressing, and use of geodetic data.

Spatial metadata is provided through the ASDD, and forms a key part of the ASDI. The ASDD provides the community with information about existing datasets and the limitations on the use of those datasets. To participate as a node of the ASDD, an organisation must meet the following criteria (Hall 2002):

- manage a collection of geospatial dataset descriptions that comply with the ANZLIC Metadata Guidelines and the most recent ANZMETA XML Document Type Definition (DTD);

- dataset descriptions must be complete and meaningful, and preferably include scale and resolution of data;

- dataset descriptions must be made freely available at no cost;

- run a Z39.50 (an international standard for communication between computer systems) server that responds correctly;

- the node administrator must have authority from the data custodians to publish the data. The data should not be listed on other nodes;

- the node must be available to all network clients; and

- registration and description of the node must be complete.

Licensing, distribution and dissemination

Although spatial data is freely available, all data is accompanied by a licence when transferred. Licence arrangements are required to ensure that spatial information is accessible, while protecting copyright, intellectual property, privacy and confidentiality. The licence also requires that the rights of the individual and governments in relation to confidentiality, privacy, security and intellectual property must be preserved, and that the rights of all parties are protected and understood.

The ASDD provides information about available datasets, whilst the actual dissemination and distribution of information is controlled by individual agencies.

ANZLIC is currently developing a 'clearing house' mechanism for the distribution of its spatial data, in order to make access quicker and easier for those who wish to use the data. ANZLIC has conducted a workshop which resulted in the establishment of a Clearing House Working Group to make recommendations on future developments of spatial data access.

Pricing policy

The pricing policy of ANZLIC and the Australian government is that all fundamental spatial data should be available at marginal cost of transfer, in order to maximise the net economic and social benefits arising from its use. Marginal costs are considered to be the costs of physical dissemination, such as those of media, extraction, packing, postage and distribution. However, agencies are currently making spatial data available through their websites, a policy that is consistent with government online initiatives. Datasets distributed through the Internet have virtually no marginal costs, so ultimately all fundamental spatial data will be available free of charge.

Archiving policy

All data produced under ANZLIC is archived unless this is prevented due to commercial, confidentiality, copyright or contractual arrangements. The intention of ANZLIC's archiving policy is to ensure that data is available for uses such as ongoing monitoring and natural resource assessments. Archived data is time stamped and version controlled, and is not changed, amended or altered unless this is necessary to correct an error during the archiving process.

In New Zealand, there are specific arrangements made for archiving. The government department known as Archives New Zealand manages the implementation of the Public Records Act and is responsible for working with the heads of all other government departments to agree what information to keep and what information to discard. This approach is taken for all government information, for example shipping and health data, and is certainly applicable to spatial information. If a government department wishes to dispose of a dataset then it must act within a legal framework and follow a set procedure to effect the disposal.

Cabinet level view

In August 2000 the Australian Cabinet established an Interdepartmental Committee on Spatial Data Access and Pricing (IDC 2001). The stimulus for the establishment of the committee was the realisation that fundamental spatial information provides part of the national infrastructure in a similar way to roads and railways. In Australia, the annual spend on fundamental spatial data has recently been approximately AU$200 million by the federal government plus a similar annual total by state and territory governments. So the Australian Interdepartmental Committee saw that there were strategic and financial benefits in improving access to spatial data and produced the four main recommendations reproduced below (IDC 2001). In the recommendations the use of the word *Commonwealth* refers to the Commonwealth of Australia:

1   Provide fundamental spatial data free of charge over the Internet, and at no more than the marginal cost of transfer for packaged products and full cost of transfer for customised services, without any copyright licence restrictions on commercial value-adding. Fundamental spatial datasets … will be identified in a public schedule.

2   Develop an Internet-based public access system, within the framework of the Australian Spatial Data Infrastructure. Agencies will be responsible for maintaining their own data access and management systems, but must comply with an agreed set of standards.

3   Negotiate a multilateral agreement with the [Australian] States and Territories for access to spatial datasets required for Commonwealth purposes.

4   Replace the Commonwealth Spatial Data Committee with a new administrative structure, comprising an executive policy group, a management committee, and a new Office of Spatial Data Management. This structure will be responsible for managing the new access and pricing policies.

The new Office of Spatial Data Management has developed a corporate plan to expedite the delivery of spatial data, information and knowledge for the economic, social and environmental benefit of Australia.

The attacks on New York on 11 September 2001, along with subsequent terrorist attacks in Bali, Madrid and elsewhere, have raised the profile of emergency service responses. Stimulated by the threat of terrorism, Australia has launched an initiative on counter-terrorism in which spatial data play a key role. The initiative is the Geospatial Emergency Information Network (GEIN). It provides a database for military and policing activities, but also applies to civil emergencies such as bush fires, flood mitigation and earthquake damage. In the process, GEIN will help to drive data integration and the use of standards and metadata in an operational environment.

CANADA

GeoConnections

Canada operates a similar nation-wide policy towards geospatial data to that of Australia, with the aim of developing the Canadian Geospatial Data Infrastructure (CGDI). The lead organisation responsible for the CGDI is the Interagency Committee for Geomatics (IACG), which consists of the Canadian Council of Geomatics and a number of federal government agencies. The scheme is being implemented by GeoConnections, a national collaborative programme whose aim is to develop the CGDI and make Canada's geographic information available on the Internet. This is implemented through a number of 'nodes' within GeoConnections.

Data policy issues are the responsibility of the GeoConnections Policy Advisory Node, which assesses policies related to data access and use, identifies data licensing and distribution issues, and promotes the harmonisation of data policy issues to facilitate broader data sharing. The Policy Advisory Node has been responsible for developing a guide to best practice in dissemination of government spatial data in Canada (Wershler and Rancourt 2003). At the start of the guide the authors note an imbalance over access to data that is common in many countries:

> … the data dissemination and licensing frameworks used to promote, extend and support the use of government geographic data generally have not kept pace with developments in technical capacity and growing user demand … The variety of terms of use, fee structures, source acknowledgment and termination clauses used by federal departments makes it difficult to optimise the use of government geographic data.

The guide clarifies the laws that govern access to data in Canada, and proposes three models of dissemination of government data: unrestricted use licence; end-user licence; and a distributor licence. Each model has benefits to government departments, and the guide as a whole is designed to help the data suppliers to be more explicit over roles and responsibilities in disseminating spatial data, especially over the liability conditions. The guide includes example or template licence agreements that can be used by government departments.

Ownership, privacy and confidentiality

GeoConnections takes a similar view to ANZLIC with regard to the ownership of core spatial datasets, and considers them to be a public good that should be freely available. In treating this core framework as a public good, GeoConnections encourages wide use, consistency and standardisation. Digital spatial data created by the Canadian government is only restricted if it compromises privacy, or if it is classified as secret for reasons of national security.

Intellectual property rights and associated legal frameworks

Broader use of geospatial data in Canada has been hindered in the past due to the restrictive Crown copyright requirements that prevented the redistribution of data. GeoConnections now adopts the policy that instead of preventing data use, licensing and copyright should be used to protect data integrity, essentially building a 'branding' to exploit the credibility of the original source data (Sears 2001). This essentially means that users may redistribute datasets as long as the original source is acknowledged and the data is not altered. Such a policy is aimed at maximising data use and the benefits that result.

Standards and metadata

As in the case of Australia, the CGDI is being developed in accordance with internationally recognised technical standards and specifications that enable interoperability between the agencies and users wishing to access CGDI data. Endorsed standards are developed through the Open GIS Consortium (OGC) and the International Standards Organisation Technical Committee 211 on Geomatics.

The policy towards the provision of metadata is also similar to that of Australia, and is seen as key to the implementation of the CGDI. All datasets must have well-documented descriptions of their availability, content and applications. In addition, users can search for metadata from a wide variety of sources using the Data Discovery Portal located on the GeoConnections website (GeoConnections 2004). Whilst this does not supply data, it describes the location, availability and coverage of a large number of geospatial datasets.

Licensing, distribution and dissemination

While GeoConnections aims to distribute geospatial data as widely as possible, licensing is seen as an important way to protect the quality and integrity of government geospatial data. Licences are seen to have a number of benefits, such as documenting rights and responsibilities, facilitating partnership and business development and advancing agency recognition or branding. However, GeoConnections recognises that licensing can both restrict and assist data access, and GeoConnections therefore aims to harmonise licensing policies. This includes clarifying policies pertaining to intellectual property and harmonising redistribution licences.

National framework data is disseminated through the GeoConnections website (GeoConnections 2004) and can be obtained free of charge. Other geospatial data can also be located using the Discovery Portal, a search engine that locates metadata for a large number of spatial datasets. The aim of this system is to help people not only find data but also tools and web services that will help them to develop geospatial applications.

Pricing policy

Core framework geospatial data is regarded as a public good, and is provided free of charge via the Internet to encourage use, consistency and standardisation. Costs for private benefits, which are seen to be beyond the public good, are borne by the user. Whilst data provided via the Internet is free of charge, GeoConnections operates a pricing model to facilitate efficiency. Essentially, this means that data not provided electronically is provided at the marginal cost of reproduction or at 'nuisance fees' that discourage requests for data in this format.

For other types of data, the government operates a cost-recovery policy, where some or all of the costs of a government activity are transferred from the taxpayer to those who directly benefit from the service. Reasonable costs are also recovered from clients when the government provides a value-added service, although these cases are limited to those services that cannot be provided by the public sector.

An example of a change in attitude on pricing policy is the National Atlas of Canada at 1:1 million map scale. At first, there was a charge for the atlas that represented a cost-recovery approach. However, the revenue per annum from sales of the National Atlas was C$45,000 while the transaction cost of administering the sales was C$75,000 per annum. So, it was costly for the Canadian government to operate in cost-recovery mode for its National Atlas. As a consequence, the National Atlas is now available for free over the Internet.

Archiving policy

Archiving policy is currently being developed by the Archiving Working Group within GeoConnections, and Canada is addressing the problems surrounding long-term preservation of digital geospatial data. GeoConnections believes that data custodians should be responsible for data management plans that include completeness of data inventories and data archiving.

UNITED STATES

Characteristics

Like Australia, New Zealand and Canada, the US is regarded as a leader in the National Spatial Data Infrastructure (NSDI) field, and follows policies that ensure data produced with public money is widely available. In April 1994, President Clinton signed Executive Order 12906, which called for the establishment of the US NSDI. The motivation behind the NSDI was that a consistent, reliable means to share geospatial data among all users would result in significant savings for data collection, enhanced use of data and better decision-making. The Office of Management and Budget issued a revised Circular A-16 that reaffirmed the US government's commitment to building the NSDI in August 2002. This section reviews the US position on data policy in relation to its national spatial data infrastructure.

Ownership, privacy and confidentiality

The US Federal Geographic Data Committee (FGDC) is responsible for co-ordinating the development of a US NSDI. The NSDI encompasses policies, standards and procedures for organisations to co-operatively produce and share geographic data. Data produced under publicly-funded agencies is regarded as being publicly-owned, and is therefore made as widely available as possible under the FGDC.

Intellectual property rights and associated legal frameworks

IPR regarding federally produced data are unrestrictive and data is viewed as a national resource in the US. Consequently, the unrestricted sharing and redistribution of geographic data is encouraged in order to maximise the return on the investment of public resources. In some cases, selected principal investigators have initial periods of exclusive data use, but the data is then made openly available as soon as the exclusive use period has expired.

Standards and metadata

FGDC standards are developed through a structured, open consensus process and are integrated with one another and with voluntary consensus standards. The FGDC has formally endorsed twenty geospatial standards. These include the Spatial Data Transfer Standard, Content Standard for Digital Geospatial Metadata, Cadastral Data Content Standard, Content Standard for Digital Orthoimagery, and Geospatial Positioning Accuracy Standards.

In 1994 the FGDC adopted the Content Standard for Digital Geospatial Metadata to label or document geospatial datasets. The standard provides a common set of terminology and definitions for the documentation of geospatial data, including data elements covering the following topics:

- identification information, such as title, geographic area and scope of validity;

- data quality information, such as positional and attribute accuracy, completeness and consistency;

- spatial data organisation information, including the method used to represent spatial positions and the number of spatial objects in the dataset;

- spatial reference information, for example the name of and parameters used for map projections or grid co-ordinate systems;

- entity and attribute information, such as the name and definitions of features, attributes and attribute values;

- distribution information, such as contact information for the distributor, available formats and information about how to obtain the data; and

- multi-use sections, which may include the time period covered by the dataset, citation information and information sources from which the data was derived.

Licensing, distribution and dissemination

The FGDC aims to disseminate geographic data in a manner that achieves the best balance among the goals of maximising the usefulness of the data and minimising the cost to the government and the public. Data products should be disseminated equitably and on timely and equal terms.

The National Geospatial Data Clearinghouse is the means of finding geospatial data from a variety of governmental and non-governmental sources, determining their fitness for use, and identifying the means for obtaining, accessing or ordering data as economically as possible. The Clearinghouse is described as neither a central repository where datasets are stored, nor a set of websites referencing spatial data, but a federated system of compatible geospatial data catalogues. Metadata is made available through the Clearinghouse so that the government and the public can determine what geospatial data exists, the condition of the data and how to access it. Each data producer is expected to describe available data in electronic form and prepare the metadata for the Clearinghouse using either free or commercially supported software tools. One of the essential requirements of the Clearinghouse is to support the search for geospatial data over the Internet. Metadata is managed in collections maintained by each organisation and made searchable by full-text and fields.

Pricing policy

Agencies set use charges for data products at a level sufficient to recover the cost of dissemination, but no higher. The charge also excludes costs associated with the original collection and processing of the data.

Archiving policy

The National Archives and Records Administration (NARA) is the federal agency responsible for acquiring, preserving and making available those records of enduring value created or received by the US federal government. Federal agencies are required to manage records in accordance with NARA regulations. These require the approval of the Archivist of the United States before any Federal Records are destroyed, stored in Federal Record Centres, or transferred to the National Archives for permanent preservation.

Records in geospatial database systems that provide evidence of the organisation, policies, programmes, decisions, procedures, operations or other activities of an agency of the federal government may be deemed appropriate for preservation. A broader body of geospatial data may be preserved because of the value of the information it contains. The following questions are considered before the disposal of data (FGDC 2003):

- Does the data involve or reflect any legal rights of the government or individuals?

- Will the data be needed to defend the agency or the government against charges of fraud or misrepresentation?

- Could the data be useful to other federal geospatial data users or the broader research community?

- Will other users require access to the original 'raw' data?

- Have the geospatial data been made available to other users through agency data sharing agreements, data user services or the Clearinghouse?

- Can secondary users understand or interpret the data without technical expertise or assistance from the producer?

- Is the data difficult or expensive to replicate?

- Are there significant costs or consequences to the programme or the government if the data is lost?

- Can the data be usefully integrated with newer data resulting from resurveying or improved methods of data collection and interpretation?

- Does the estimated research value of the data exceed the costs to maintain them for secondary use by government researchers or others?

- Will the data be useful for analysing geographic distributions over time?

- Does the data support the study of geophysical changes over time?

If the answer to any of the above questions is 'yes', the data may have long-term or permanent value and is therefore considered for archiving.

US National Satellite Land Remote Sensing Data Archive

Establishment

In 1992 the US Congress directed the Department of the Interior to establish a permanent government archive containing satellite remote sensing data that would be easily accessible and ready for study. This was defined under Public Law 102-555, which stated:

> It is in the best interest of the United States to maintain a permanent, comprehensive Government archive of global Landsat and other remote sensing data.

The US Department of the Interior is a key player in this process and was mandated to:

> ... provide for long-term storage, maintenance and upgrading of a basic, global, land remote sensing dataset ... and shall follow reasonable archival practices to assure proper storage and preservation of the basic dataset and timely access for parties requesting data.

The location for this archive was established at the United States Geological Survey's EROS Data Center, and the collection of information is known legally as the National Satellite Land Remote Sensing Data Archive (NSLRSDA). This archive forms a comprehensive, permanent and impartial record of the land surface from almost 40 years of Earth observation data.

Efforts to address long-term data preservation and access have recently been made by remote sensing leaders from academia, industry and government as members of a federal advisory committee. An NSLRSDA advisory committee is viewed as necessary to influence the guidelines or rules relating to the selection of data for archival deposit, maintenance and preservation as well as access management policies and procedures. The committee prepares white papers, recommendations and proposed policies for the NSLRSDA. Their work as a federal advisory committee has the potential to assist the archive on a long-term course of acquiring, storing, maintaining, preserving and upgrading satellite data for the public good.

Ownership, privacy and confidentiality

Data held in the NSLRSDA is generally from publicly-funded missions, and ownership therefore ultimately resides with the US government. However, some data produced by privately operated remote sensing systems is also held in the NSLRSDA. US firms that are licensed to operate private remote sensing systems are required to make available unenhanced data requested by the NSLRSDA on reasonable cost terms as agreed by the licensee and the archive. After a period of time, the archive may make this data available to the public at a price equivalent to the cost of fulfilling user requests.

Intellectual property rights and associated legal frameworks

United States Geological Survey (USGS) authored or produced data or information is in the public domain. USGS requests that users of their data acknowledge the USGS as the source.

Licensing, distribution and dissemination

A primary objective and major challenge of the archive is to distribute data on demand to a worldwide community of scientific users. Data access is provided through the online facility known as EarthExplorer (USGS 2004). EarthExplorer functions as an online, interactive source of information on Earth observation data, and includes browse, search and order capabilities for the various datasets. Unenhanced data in the archive generated by the Landsat system or any other Earth observation system funded and owned by the US government is made available to all users on a non-discriminatory basis.

The format on which data is supplied depends on the type of product. For example, declassified satellite imagery is provided on film or as a paper print, whereas Landsat data is disseminated on CD-ROM, 8mm tape or electronically via the Internet.

Pricing policy

The USGS states that the data they provide should be available to all users at costs that the users can bear, and this should be a price equivalent to the cost of fulfilling user requests. In reality, this varies between the type of product, the level of processing and the format in which it is supplied. For example, geometrically registered data from the Advanced Very High Resolution Radiometer (AVHRR) costs US$190 whereas Level 1B data is supplied at US$50. Declassified satellite imagery from the KH1 to KH6 missions is charged at US$18 for black and white negatives, or around US$14 for paper prints. Images from the KH7 and KH9 missions are more expensive, at US$30 for black and white negatives, and US$65 for colour positives. Paper prints are also more expensive, at US$18 for black and white prints and US$40 for colour prints.

The price of Landsat data also varies according to the sensor (for example, Thematic Mapper (TM) or Multispectral Scanner (MSS)) and the level of processing. Landsat TM products are priced at US$425 plus US$200 for additional scenes, whilst the Enhanced TM products are more expensive at US$600. These prices are less for US government and affiliated users and approved researchers. MSS photographic products are priced at US$10 for black and white film, whilst digital products are US$200 for a corrected single scene.

Archiving policy

The collection of data from Landsats 1 through 5 forms the core of the national archive, in addition to 35,066 gigabytes of data from AVHRR and 992,674 declassified intelligence satellite photos. In addition to this data, archive holdings also include data from Landsat 7, ASTER, SRTM, LightSAR and NASA's Small Spacecraft Technology Initiative. The total projected holdings by the year 2005 are estimated to be 1,400,000 gigabytes of data (USGS 2003).

The EROS Data Center currently uses various technologies to archive its offline datasets. In 1992, the TMACS (TM/MSS Archive Conversion System) was used to transcribe Landsat data archives from High Density Tape (HDT) to Digital Cassette Tape (DCT). Both these formats utilise large, expensive analogue drives that make the cost of transcribing HDT to DCT exceed US$1 million for each generation of media. The ongoing maintenance costs of HDT and DCT drives were found to be excessive as there is little industry experience and only a single vendor to support each brand of drive. There are only a few HDT and DCT drives left in existence, with the number of drives decreasing each year as other users migrate to digital media.

The USGS has determined a series of criteria that should be used to determine archiving methods and media (USGS 2003):

- Technology must be currently available and the latest drive in the line.

- The technology must hold more than 50GB of data.

- The technology must have a fast write transfer rate.

- The technology must use media that can remain readable for at least 10 years in a controlled environment. Ten years was deemed an appropriate length of time since it is the longest that a media technology would conceivably be used before space and transfer rate concerns would dictate a move to new technology.

- The technology must not be hampered by a poor reliability history.

CONCLUSIONS

One conclusion from the experience of spatial data infrastructure development in Australia, New Zealand, Canada and the US is the recognition of the need to be explicit in stating the conditions of access to data and information. Laws and policies are being developed and used to clarify the conditions of access to data, and this increasing clarity is assisting in the maturity of the spatial data sector. The concept of a data infrastructure being as significant as a road infrastructure is a powerful one. Europe is lagging behind, yet the Global Monitoring for Environment and Security (GMES) initiative can assist Europe in making substantial steps rapidly in developing and using better and more explicit policies that are clear on the roles, rights and responsibilities of environmental data suppliers, data users and regulators.

CHAPTER 7

DATA ACCESS AND THE INTERNET

INTRODUCTION

The rapid uptake and development of the Internet on a global basis in recent years has meant that it has become increasingly easy in principle to obtain and distribute data and information. From a global monitoring perspective, this offers the potential for increased data sharing and use, widespread dissemination of new scientific findings and wider discussion and analysis of new ideas. In addition, government initiatives designed to increase freedom of information and citizen participation have led to the development of e-government websites and data infrastructures, allowing wider access to government data and information. The Internet is often regarded as an ideal means for distributing data on an egalitarian basis to any interested parties, allowing widespread access at minimal cost. However, in reality the economic development of a country often governs its level of Internet connectivity, resulting in disparity between rich and poor in the 'information economy'. This chapter discusses the increasing use of the Internet for data dissemination through the use of data portals and web services, and the implications of variable access to the Internet for less economically developed regions of the world.

DATA PORTALS

Introduction

Many governments and public bodies are increasingly keen to publish the data and information they produce in order to increase transparency and maximise the usage of publicly-funded data and information. Consequently, concepts such as e-government and spatial data infrastructures have led to an increase in the use and development of online data portals and web services to distribute data and information to interested parties. Similarly, data portals have proved to be advantageous for groups of users or organisations that require access to the same basic datasets for similar functions. Shared data and functionality ensures consistency, allowing data integration and comparability, maximising data use and reducing the inefficiency of a single dataset being recorded more than once by different users.

The development of data portals can also be partly attributed to the widespread adoption of certain standards, such as the use of Extensible Markup Language (XML) (Tang and Selwood 2003). The benefits and widespread use of

XML have meant that organisations have increasingly based their own standards on XML. For example, the Geographic Markup Language developed by the Open GIS Consortium (OGC) is an XML schema designed to allow compatibility between spatial datasets.

The Geography Network

The Geography Network (2004) is essentially a web-based catalogue or registry of Internet services that acts as a means of sharing and disseminating spatial data. The network is managed and maintained by the ESRI company, and provides a single point of access to a wide range of data and information that is freely available to any organisation that wishes to participate.

Essentially, the network acts as a single source for users who wish to discover and use spatial datasets. Instead of having to search through the catalogues of individual data providers, the Geography Network provides a 'one-stop shop' for locating spatial data and provides the users with information about the data held at many servers. In many cases, datasets can be viewed online, downloaded or integrated with a user's GIS system. The Geography Network supplies three categories of geographic content: data, documents and resources.

Users can search and preview metadata and geographic information using the Geography Network Explorer. Using this interface, users can search the Network using criteria such as location or data type, or can browse through the metadata organised into a hierarchy of directories. The results are then summarised with relevant descriptions, including the name of the publisher, content title, coverage area and map scale. Users can then select the records to see their detailed description, or preview the information on a map. Most of the data can then be downloaded or directly added to Geographic Information System (GIS) software packages and applications.

E-government and data portals in the UK

The agenda for e-government has been highly publicised in the UK, and much has been made of its potential to increase participation in civic and political issues. It has been said that e-government schemes, if implemented properly, can improve government services and the delivery of services, increase accountability, reduce administrative costs and time spent on repetitive tasks for government employees, facilitate greater transparency in government administration, and allow greater access to services through around the clock availability on the Internet (Jaeger 2003). Consequently, ambitious UK targets, outlined in the *Modernising Government* White Paper of 1999 and the e-Government Strategic Framework of 2000, aim to introduce 100% online services by 2005.

There are a number of high profile examples of data portals for the distribution of UK government data and information. Maps Direct (2004) is one such example, and provides services to public and commercial organisations. Maps Direct is the branding for a suite of Internet-based services that are hosted by ESRI (UK): its aim is to provide a mapping tool as a single licence package which delivers up-to-date, actively maintained map data that can be used throughout an organisation and can be integrated with existing applications. An agreement between the UK government and Ordnance Survey in April 2002 means that Maps Direct is able to provide access to key Ordnance Survey datasets for 550 government departments under a single licensing regime (Tang and Selwood 2003).

The Planning Portal (2004) is another e-government initiative, and acts as a first point of contact for anyone who requires information about the planning system in England and Wales. The portal aims to provide information to a wide range of users, including government departments, planning professionals and the general public.

The GI Gateway (AGI 2004) is a portal that assists users in locating and using up-to-date and accurate UK geographic information. The service is funded by the government through the National Interest Mapping Services Agreement (NIMSA) and aims to increase awareness of and access to geographical information in the UK. NIMSA is an agreement between the Office of the Deputy Prime Minister and Ordnance Survey to provide funds for mapping activities that cannot be justified on purely commercial grounds.

Global Environmental Outlook data portal

The Global Environmental Outlook data portal (UNEP 2004) is maintained by the United Nations Environment Programme (UNEP), and holds more than 400 variables as national, sub-regional, regional and global statistics or as geospatial datasets. The data covers a variety of environmental themes, such as freshwater, population, forests, emissions, climate and a number of socio-economic themes, including GDP and health.

The portal allows the user to search the database using a keyword, for example 'earthquake' or 'population', and a list of matching datasets is then displayed. The list also describes the type, time period and region of the dataset. For example, entering the search term 'earthquake' reveals a list of datasets that includes a geospatial dataset of earthquake intensity zones, datasets containing the numbers of people affected or killed by earthquakes, and datasets detailing the number of earthquakes on national, sub-national and regional scales. The user can select one of these datasets and the year in which they are interested. The data can then be viewed as a map, graph or as a table of values, or can be downloaded in PDF, Microsoft Excel or ESRI Shapefile formats. In a limited number of cases, downloading the data is controlled by password protection to UNEP collaborating centres, and is available for internal institutional use only.

The UNEP website also provides information content through a number of specific thematic portals, such as climate change and socio-economic factors. Much of this content is supplied as reports, articles and interactive resources rather than raw data, and is often held by third parties such as the World Bank, the World Trade Organization (WTO) or by other UN agencies. The UNEP site provides metadata and links to this information so that interested users can assess its suitability for their purposes, and find out how to access the information. This information is not always available free of charge, and some of the organisations listed levy a fee for certain datasets or for printed publications.

GeoConnections data portal

The GeoConnections data portal (GeoConnections 2004) represents the access point for data and information provided through the Canadian Spatial Data Infrastructure, and aims to make Canadian geospatial data, tools and services readily available online. The portal provides access to maps and images, including navigational charts, diagrams, digital elevation models, political maps, paper maps, digital raster data and remote sensing data. In addition, users can search for online mapping tools and services.

Users can also search for geomatic datasets by Canadian Province or Territory, by thematic data types and subject, or for framework geospatial data such as ecological units, roads and railways. When a user has defined the type of information they require, the relevant datasets are listed. Symbols next to each listed dataset indicate whether the data is Canadian, if it has been quality assured, if it is provided free of charge or if it is searchable online. Most of the datasets listed cannot be directly downloaded from the GeoConnections data portal, but the user is provided with detailed metadata that describes the dataset origin, abstract, purpose, information about the supplier, thematic coverage, where the data can be obtained and if it is available for free. The GeoConnections portal also provides a list of other Canadian and international sources of geospatial data and services.

VARIATIONS IN INTERNET CAPABILITY

Introduction

The provision of timely and accessible data is key to the successful monitoring of the global environment. Rapid access to data is particularly important in relation to international security, in order to direct aid and recovery operations after a disaster, for example. The development of data portals, e-government initiatives and spatial data infrastructures illustrates how the Internet is becoming one of the most frequently used methods for data dissemination, allowing easy access to data wherever a computer connection is present. Data dissemination via the Internet is already widespread, particularly for government-funded information and data produced in the public interest. Such policies are designed to increase data sharing and allow equal access to data and information. Whilst this is true for more economically developed countries where many people have access to the Internet, the same cannot be said for less economically developed countries where Internet access may be limited and more expensive relative to earnings.

Current situation

International connectivity to the Internet has increased dramatically in the last 10 years, so that by 1997 most nations had some form of connection. The exceptions are generally nations suffering from extreme poverty, war and civil conflicts or from geopolitical isolation (Dodge and Kitchin 2001). The diffusion of the Internet throughout the last decade is illustrated in Figures 7.1 and 7.2.

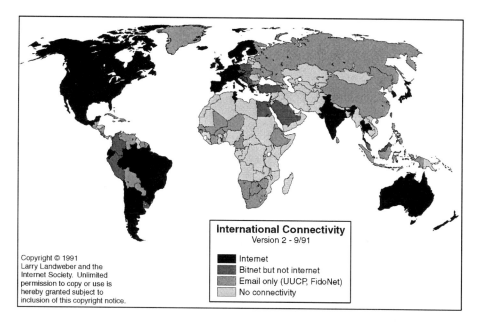

Figure 7.1: International Internet connectivity in 1991. Source: Landweber (1991)

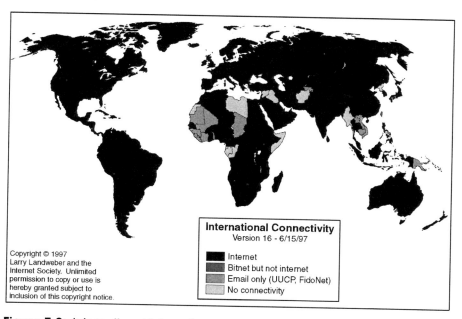

Figure 7.2: International Internet connectivity in 1997. Source: Landweber (1997)

However, whilst most countries are now connected to the Internet, the level of connectivity is not equally distributed in scope or cost, and there is a disparity between the more economically developed and the less economically developed regions of the globe. Africa has the most obvious lack of Internet hosts, in stark contrast with Europe and North America. Internet penetration as a percentage of population is as low as 1.1% in Africa, compared to 66.1% in North America and 28% in Europe (Internet World Statistics 2004). A major contributing factor to the lack of Internet availability has been the severe lack of an adequate telecommunications infrastructure in many developing countries, particularly in Africa. It has been reported that outside of Nigeria and South Africa, only 1.9 million people in sub-Saharan Africa experience a regular phone service. Also, most of the telecommunications infrastructure that does exist in Africa cannot reach the majority of the population, with 50% of the available lines concentrated in the capital cities where only about 10% of the population live. Wired telecommunications networks are unlikely to develop significantly in the near future, because of the increase in mobile and satellite communications technologies. Telecommunications services are also more expensive in Africa, with the cost of renting a connection averaging almost 20% of GDP per capita, compared to a world average of 9% and only 1% in high-income countries. The poor network connections are highlighted by the significant disparity between outgoing bandwidth from African countries that is destined for the US and Europe and that which is intra-African. Of the 1.5 Gbps of outgoing bandwidth, approximately 1 Gbps lands in the US, 375 Mbps in Europe and only 13 Mbps is intra-African (Jennings 2002). Internet traffic within Africa is often sent via Europe and North America, a process that is both inefficient and expensive, particularly as developing countries with weak currencies have to pay in US dollars or euros for most of their upstream bandwidth.

Such problems mean that bandwidth in Africa and other developing countries can be extremely expensive. A report by the International Network for the Availability of Scientific Publications (INASP) revealed that most universities in the developing world cannot afford more than 1.54 Mbps of bandwidth, a figure that is particularly low when compared to the 2.5 Gbps experienced by many British universities (INASP 2003). It has been suggested that if the value of real estate is dependent on location, then the value of a network connection is determined by bandwidth, with accessibility redefined so that the 'friction of distance' is replaced by the 'bondage of bandwidth' (Mitchel 1995, cited in Dodge and Kitchin 2001). High international tariffs and a lack of circuit capacity means that obtaining sufficient international bandwidth is still a major problem in most African countries, and users generally have to contend with substantial congestion at peak times. Figure 7.3 illustrates the differences in bandwidth between more and less economically developed regions of the globe. Many fear that developing nations will become increasingly marginalised from the information or knowledge economy if bandwidth remains limited and expensive.

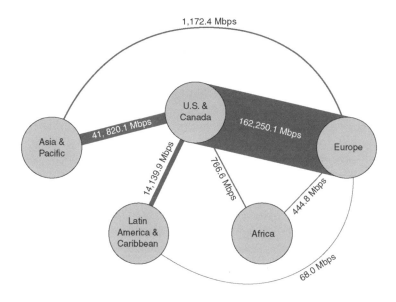

Figure 7.3: International Internet bandwidth, 2001. Source: Global Internet Geography (2004)

Mbarika *et al* (2002) summarise the key reasons why Internet development in African countries has been inhibited:

- Internet services are limited to the urban centres in many countries, with very few Post Office Protocol (POP) centres located outside of major cities. This is mainly due to poor wired telephone connections in rural regions.

- Problems with local telephone infrastructure, such as low levels of teledensity, low-speed and narrow bandwidth lines, poor-quality connections and high land-line telephone installation and usage fees.

- The high cost of available international long-haul links to high-speed Internet backbones, which usually translates into high subscription and connection charges for end users.

- The acute shortage of technical staff to design, install, operate and maintain networks and Internet technologies is a serious impediment to growth. A shortage of properly skilled people has meant that many Internet initiatives have been delayed or postponed. Robison and Crenshaw (2002) attempted to model the global spread of the Internet using a number of macro-social variables. One of their most important observations was that educational attainment (measured as secondary school enrolment ratios in 1985) conditions the influences of the other variables in their model.

Implications of Internet connectivity variations and access to environmental information

Global variations in Internet accessibility and bandwidth have a number of implications for accessing environmental information. The Internet is a widely used method of dissemination for environmental data because data distribution can be maximised and marginal costs can be kept low. This is clearly the most accessible way of obtaining data for users in Europe, the US and other regions where many people have access to the Internet with high bandwidths. However, for users in less economically developed regions such as Africa, where Internet access is limited and expensive, obtaining the data can be considerably more difficult. Even where Internet access is present in these regions, the low bandwidth is likely to mean that downloading large datasets is extremely time-consuming and expensive. This could mean that aid agencies in areas that require the data in an emergency have difficulty gaining access, for example in times of drought or famine. Similarly, regular environmental monitoring programmes in Africa may be hindered by a lack of easily accessible datasets.

A number of programmes within Europe are currently aiming to produce data that will be potentially useful for aid agencies and governments in Africa. For example, the Global Monitoring for Food Security project (GMFS 2004) aims to provide advance crop information derived from Earth observation data, focusing primarily on sub-Saharan Africa. The information will be used by existing regional, national and international early warning systems that monitor agriculture for food security, as well as international aid organisations. Similarly, the Joint Research Centre (JRC) has developed a number of programmes to provide data for external aid and security policies that will directly benefit African nations. For example, an online system known as the Digital Map Archive (JRC 2004) has been developed to supply geographic information, such as maps and satellite imagery, to support the work of the European Commission's humanitarian aid office. The information can be used to help locate crisis areas and identify access routes so that planners and aid agencies can assess and react quickly to a crisis. This information is currently used internally in the European Commission to develop external aid and security policies. African governments and aid agencies within Africa are likely to benefit from direct access to this type of information. However, the limitations of Internet access mean that downloading and accessing such information electronically within Africa is time-consuming and expensive.

The Ptolemy Project provides an example of how providing data and information via the Internet has significant benefits for developing nations (Beveridge *et al* 2003). Doctors in Africa have very limited access to the journals and texts that are viewed as fundamental to good practice and medical research. The aim of the Ptolemy Project was to provide access to electronic health information and to analyse whether that service had a positive effect on surgical research, teaching and practice in East Africa. The results of the Ptolemy Project indicated that the provision of such information had a significant positive impact on the medical research and practice of the respondents, and elicited numerous positive comments. Increased access to environmental information could have similar beneficial effects on environmental research and civil security in Africa as those seen in the medical sector under the Ptolemy Project.

The Ptolemy Project also provides examples of the problems caused by Internet access in Africa. For example, the commonest criticism of the utility of electronic health information in the developing world relates to slow, unreliable and costly or non-existent Internet access. Nearly all the respondents complained about their Internet connections, yet 61% browsed Ptolemy for more than an hour a week, and 68% estimated their total combined Internet and telephone costs at around US$50 a month (Beveridge *et al* 2003). This represents a considerable amount of money for a physician in Africa, highlighting the problems of limited and expensive bandwidth.

Improvements in Internet access

In order to solve the problems associated with the disparities in the technology, there have been several projects that have aimed to improve the international telecommunications network, particularly with regards to Africa. As a result, recent reports by the International Telecommunication Union have indicated that lack of infrastructure is no longer the limiting factor in extending Internet access and bandwidth. However, other factors such as affordability of access, level of education and quality of services are becoming more significant (*The Economist*, 29 November 2003).

Telecommunication executives from East Africa have discussed in recent years the possibility of connecting countries in the region by an undersea fibre-optic cable, possibly with a strategic partner from outside Africa. South African and West African firms were co-opted to lay an undersea link for the west coast of Africa without outside investment, and there are currently five submarine fibre-optic cables providing some connectivity to Africa. These cables connect most of the North African and West African coastal countries from South Africa to Morocco to Europe (Jennings 2002). These improvements in infrastructure mean that much faster Internet connections are now possible within Africa (see Figure 7.4 overleaf). However, these marine cables often do not carry enough Internet traffic to achieve the economies of scale that makes bandwidth in other parts of the world so affordable, and bandwidth therefore continues to be expensive (INASP 2003).

Where physical cable connections are not possible, Internet connections can be made via Very Small Aperture Terminal (VSAT) satellite connections. Satellite providers such as Intelsat, New Skies and Panamsat provide much of the international bandwidth via satellite broadcast circuits, a common response to the bandwidth problem in which data broadcasting services are now being installed by African internet service providers (Jennings 2002). A basic satellite dish is used to receive a stream of web data for caching locally, and can provide incoming bandwidth in chunks of 64 Kbps for about US$30–$1,000 per month (Jennings 2002). This is usually more affordable than services provided by local operators, but much slower and more expensive when compared to high-speed leased-line and fibre-optic connections. One example of a system that uses satellites for networking is Healthnet, an email-based communication system that is used for epidemiological data exchange. The Healthnet system uses low Earth orbit satellites, simple ground stations and radio- and telephone-based computer

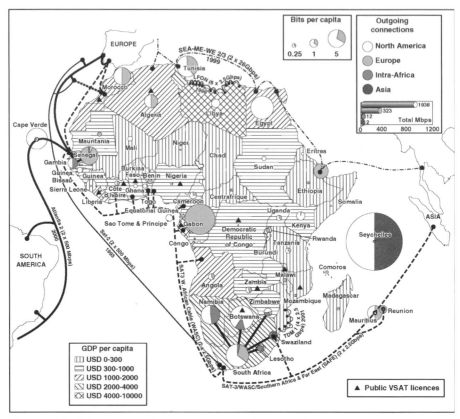

Figure 7.4: Telecommunications accessibility in Africa: GDP, bits per capita, connectivity and telecommunications cables in Africa. Source: International Development Research Centre (2004)

networks to distribute data. The system is reported to function reliably and inexpensively even in areas with little or no telecommunications infrastructure (Mutula 2002).

The use of Digital Video Broadcast (DVB) systems is increasingly being used as a method of disseminating data and for providing higher Internet connections. The TV model of one uplink with multiple receivers is ideally suited to environmental data dissemination, and a number of existing systems employ DVB technology as a mechanism to distribute data. For example, EUMETSAT's Data Distribution System (EUMETCast) utilises the services of a satellite operator and telecommunications provider to distribute data files using DVB to an audience that falls within its coverage zone. This zone includes Europe and all co-operating Member States of the EUMETSAT programme. Figure 7.5 shows the main distribution area for MSG-1 data dissemination by EUMETSAT. The extension of the system to include coverage of the African continent is being initiated and dissemination trials are currently underway. EUMETSAT claims that one of the key strengths of the system is that all the data available can be received

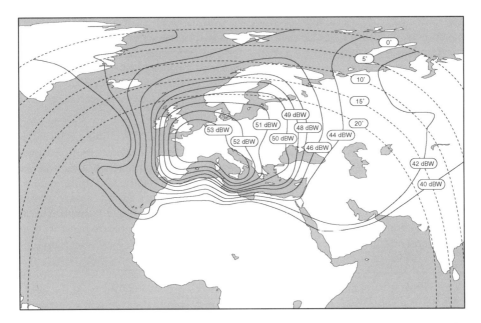

Figure 7.5: Areas of coverage for MSG-1 data dissemination. Source: Remote Imaging Group (2004)

with a single EUMETCast reception system, which greatly simplifies the required user infrastructure, and the system is significantly cheaper than the specifically developed systems traditionally used.

A DVB data distribution system has a number of advantages for environmental data distribution. For example, wide deployment of DVB technology in the cable and satellite television industries means that equipment is available from multiple sources at consumer price levels. Another advantage is that encryption and security methods currently employed by the TV industry to protect revenue can be applied to DVB data transmissions, and therefore offers a method of data encryption for restricted data. Data sent via two-way satellite Internet is in compressed format, so that the size of the data and the bandwidth are reduced. This 'multicasting' technology is capable of sending up to 5,000 channels of communication simultaneously, a magnitude that is impossible with normal dial-up technology.

SUMMARY

The Internet is widely regarded as being the most accessible and easy way to disseminate data and information. The development of web services, data portals and e-government initiatives has meant that more environmental and spatial data of a better quality is available to more people. Data portals can be used to ensure that the data they publish is produced to certain standards, ensuring that datasets are more likely to be reliable. Many data portals, such as the GeoConnections portal, have facilities to locate data and information from a variety of sources and list detailed descriptions of their attributes and functions. These initiatives have made it much easier for users and researchers to locate and access data and information for global monitoring, and ensure that this information is suitable for their needs.

Whilst this is true in more developed countries, access to information distributed via the web is more difficult and expensive in developing regions of the world such as Africa. This has been partly due to the poor telecommunications infrastructure that exists in many parts of the world, which reaches low numbers of the population and lacks connections with high capacity fibre-optic cables. While telecommunications networks have improved in recent years, and a number of new marine fibre-optic cables have been laid around parts of Africa, the cost of bandwidth in these countries remains high. This is partly due to a lack of demand and low traffic flow for the cable capacity that does exist, meaning that the economies of scale that make high bandwidth affordable in other parts of the world do not exist. In addition, there is a distinct lack of inter-African country links, which means that all Internet traffic has to be routed via Europe or the USA, which is particularly expensive when a weaker currency has to pay for upstream bandwidth in dollars or euros. Internet access in many African countries is often only available through VSAT licences, but these are slower and more expensive than fibre-optic and leased-line connections.

This has implications for global monitoring. There are a number of programmes within Europe producing data that could be extremely useful for civil security issues in developing countries, such as the JRC's Digital Map Archive and the Global Monitoring for Food Security programme. However, such data are likely to be expensive and difficult to download in the areas of the world in which they are needed the most.

One possible method of data distribution that could be used to overcome the limitations of Internet access and bandwidth is DVB technology which offers high bandwidth Internet access to all users who fall within the coverage zone of a satellite. DVB equipment and infrastructure have already been widely deployed by cable and satellite television industries, and encryption and encoding mechanisms currently employed by television networks can be applied to protecting data distribution. The EUMETCast system employed by EUMETSAT for distributing satellite data provides an operational example of such a system, and coverage is currently being extended to include the African continent.

CHAPTER 8

INTEROPERABILITY, LINKAGE AND DATA ACCESS

David Briggs, Peter Ryder and Barry Wyatt

INTRODUCTION AND METHODOLOGY

Interoperability

Interoperability is vital to the effective use of information: it can be defined as the ability to transfer data freely and seamlessly between different information systems. It brings different data together in a consistent form, which allows different models and analytical systems to link dynamically in order to derive and deliver information that would otherwise be difficult to acquire and use. Successful interoperability rests upon the establishment of:

- measurement and monitoring standards;

- data and exchange standards;

- sound methods and arrangements for linkage between different data sources and technologies (for example, Earth observation, ground-based and management-derived data) and systems for generating information; and

- suitable arrangements for integration across different sectors and policy areas, implying, for example, the linkage of environmental and non-environmental data.

This chapter is organised around these four characteristics. Before examining each characteristic, a case study of the interoperability barriers to marine monitoring is examined to illustrate the complex interactions involved.

Case study: barriers to marine monitoring

Barriers and obstacles to information use for policy support do not occur singly or in isolation. Many attempts to compile and use information face a series of different but interconnected problems. The experience of the Data Integration System for Marine Pollution and Water Quality (DISMAR) project (DISMAR 2004) illustrates the ways in which these difficulties combine to inhibit the use of Earth observation data for monitoring of marine oil spills and algal blooms (see also Chapter 5 of this book).

Oil spills and algal blooms may occur almost anywhere in the ocean, where relevant pollution sources exist. Currently, routine monitoring is performed mainly by ships and/or aircraft. *In situ* measurement systems using ferry boxes and stationary buoys are also gradually being built up, but large ocean areas cannot be covered sufficiently with such systems. Monitoring networks are not well-developed and need to be established across Europe. Ideally, satellites need to be used to supplement and strengthen these networks. Synthetic Aperture Radar (SAR) has particular capability in this respect. An operational monitoring system using SAR will require at least two satellites to provide daily coverage in all the European seas.

Effective oil spill monitoring requires the ability to maintain monitoring capability in the long term, and to direct attention to where it is needed most. This requires funding for regular data coverage of the key ocean areas, and the capability to produce and manage large data quantities. Current SAR data policy, however, is a major obstacle. SAR data costs are too high for the agencies responsible for oil spill monitoring, especially when data is needed for large ocean areas such as the Norwegian coast.

Near-real time information is required for monitoring of oil spills and algal blooms. This means that satellite data need to be acquired, processed and distributed to the users within a few hours. This can be done technically today for some regions, but the service chain to deliver the data is not equally well-developed in all countries.

Oil spills may occur at any time. Oil spill monitoring thus requires daily data coverage. This is not possible with current satellites, except at high latitudes. Operational monitoring services are thus reluctant to commit themselves to rely on Earth observation sources, such as SAR, and as a consequence aircraft and ships will continue to provide the main methods for monitoring oil spills for the foreseeable future.

Action to manage and prevent oil spills and algal blooms requires complex information, relating to many different aspects of the ocean and the pollution event. Modelling of behaviour and impacts also requires different datasets to be used in combination. Monitoring relies on many different technologies, including satellites, aircraft, ships and *in situ* instruments. If responses are to be rapid and effective, all these different datasets and sources need to be linked as seamlessly as possible. Interoperability is thus a crucial and limiting issue. Developments have started to provide interoperable services from several data providers, but there is a need to do much more in this direction to establish efficient links between services in different countries facing the same problems. Improved standardisation of data, improved metadata and more effective use of Geographic Information Systems (GIS) are essential to integrate the different data streams into a seamless service.

STANDARDS: MEASUREMENT AND MONITORING

Problems in relation to measurement and monitoring standards derive from many different sources. *Inter alia*, they include differences in:

• survey design, for example geographic coverage/extent, timing;

- sampling strategy, for example where monitoring sites are located, sample density;

- choice of determinants, that is, what is measured, what indicators are used; and

- choice of laboratory or measurement methods, for example instrumentation, analytical procedures.

Such problems affect almost all global environmental monitoring, to a more-or-less equal extent, although it is noteworthy that the meteorological sector has solved some of them. This has taken many years to achieve and a strong central body was established, the World Meteorological Organization (WMO), sustained by the enlightened self-interest of its members. Standards and best practice have been agreed, published and adopted by National Meteorological Services (NMSs) in order to remain 'members of the club' and reap the benefits of that membership. Notwithstanding this partial success in one sector, many examples of outstanding problems can be cited.

In the area of marine ecosystems, for example, there is a general unfilled need to bring together information from many different monitoring systems and sources, such as on non-point source pollution, fishing, climate change and exotic species introductions. Without this it remains all but impossible to establish cause-and-effect relationships with any degree of confidence, and thus marine environment policy remains dogged by untestable argument and counter-argument, and is liable to be less effective. To obtain this information requires the simultaneous monitoring of many variables at high spatial and temporal resolutions. This cannot be achieved through a piecemeal approach, in which different agencies and monitoring systems operate independently: it requires the definition, agreement and implementation of strong and common monitoring and survey standards. The principles of integrated marine monitoring of pollution and its effects is now accepted by several governments and Regional Commissions (for example, the Convention on the Protection of the Marine Environment of the North-East Atlantic (OSPAR)). Important elements of such an integrated approach include standards for marine measurements, chemical analysis and use of biomarkers across national and international agencies active in marine monitoring, building on the work of the Regional Commissions and drawing on schemes such as BEQUALM (Biological Effects Quality Assurance in Monitoring Programmes) and QUASIMEME (Quality Assurance Laboratory Performance Studies for Environmental Measurements in Marine Samples) to improve quality control of the constituent national datasets.

Similar issues arise in relation to oil spills. Despite increased effort in recent years, much existing monitoring is partial and fragmentary, so that impacts cannot always be predicted and consequences may not be followed up. Detection and monitoring of oil spills thus depend on implementing a more integrated, standardised approach, linking data not only on the spills and their sources, but also on coastal habitats, areas of special environmental, scientific and economic interest, shoreline access and resources available to deal with major incidents and their long-term impacts. As elsewhere, this information necessarily comes from a range of sources, including aircraft, satellite-based and ship-borne sensors, together with land-based monitoring and surveys. To link these effectively requires the development and application of common and compatible standards for monitoring and survey (as well as information exchange) between the different organisations involved (Oceanides 2004).

Outstanding issues of monitoring and survey standards are also significant in relation to the general area of emergency response and security, for example flooding, geophysical hazards, food security, technological hazards and humanitarian aid. In all these areas, there is a common need for reliable, localised information in a timely manner. Obtaining this information cannot be left to chance and to efforts when emergencies arise: it requires careful planning and preparation. The information involved is also extremely diverse. Many different organisations and monitoring systems therefore need to be brought together – each often operating, traditionally, within its own sphere of interest, on its own timescale and with its own culture and technologies. *Post hoc* amalgamation of the data created or needed by these various organisations is often not feasible, or only at the cost of reducing information quality and timeliness. Instead, the long-term need must be for a more integrated approach to information gathering at source. Again, a crucial requirement in this context is for improved and agreed standards for monitoring and survey. Given the absence of formal links between the various agencies concerned (some concerned with long-term preparedness, others with disaster management, and others with monitoring and survey), this presents some difficulties, and is unlikely to be achieved quickly.

In the area of nature and biodiversity, many of the same issues arise. Monitoring and survey in this area is typically undertaken by many different organisations, some at local level, some national, some operating internationally. Both the focus of interest (for example, whether on species, habitats or land cover, and whether on stock, quality or change) and the methods of monitoring (for example, using satellite data, field mapping, population counts) vary. In terms of biodiversity, however, the total picture only becomes evident when all these different perspectives can be combined. At the European level, this becomes even more important because of the differences in the design and methodology of monitoring programmes between different countries. Information with the potential to support the Habitats Directive, the Water Framework Directive or international conventions such as the Convention on Biodiversity (CBD) therefore often vary in quality and availability.

Problems of inconsistencies and weaknesses in monitoring and survey standards are not confined to areas in which the production of integrated, routine information for policy support is new. They survive also in well-established monitoring systems, such as those on air quality, soil protection and water. In all these cases, substantial differences often exist from one country or region to another in terms of the design and operation of the monitoring network: for example, in the location of monitoring stations, the choice of determinants, averaging periods, reporting intervals, and analytical methods. Within Europe, these differences are being addressed to some extent through existing legislation (for example, the Air Quality and Water Framework Directives and their offspring), which specify monitoring requirements in some detail. Further work is needed, however, to establish and implement standards and then to apply them, retrospectively, to the large volumes of existing information on which many users rely (for example, soil maps or air pollution records).

STANDARDS: DATA EXCHANGE

Whilst monitoring standards cover the procedures of data gathering in the field, data standards are also needed to cover the subsequent processing, storage and transfers of these data. Such standards must therefore cover all the elements of the information chain, including:

- the data formats (for example, data structure, georeferencing, level of aggregation) used for data storage and reporting;

- the physical media used for storage and transfer (though this is now less of a problem than in the past);

- the metadata used to describe the information and its genealogy;

- the exchange formats and procedures; and

- the methods and procedures used for analysis and reporting (for example, choice of indicators, statistical techniques and summary measures).

In contrast to the underlying inconsistencies in survey and measurement methods, these problems are to some extent amenable to correction after the event, though this is often costly and time-consuming, and is in any case only possible if reliable information on data genealogy is available. For many users, therefore, problems of data exchange and acquisition remain major impediments: at best, they slow down access to the information and increase its cost; at worst, where timely information is required they can prevent the implementation of effective action.

As with other issues of interoperability, these problems are evident across many different areas of application. They are, however, especially important in relation to those areas that require either the synthesis of data at higher levels of aggregation, or repeated processing through different models. In both cases, multiple data transfers between different organisations and information systems may be involved.

Reflecting this, one area in which the issues of data and exchange standards are especially important is climate change. *The Second Report on the Adequacy of the Global Observing Systems for Climate in Support of the UNFCCC* (WMO 2003), for example, cites the need to sustain and strengthen existing intergovernmental mechanisms for provision and use of climate data and products, as the first of four equally high priorities. It requests that:

> In particular, for the terrestrial domain, intergovernmental agencies should consider establishing a mechanism to prepare guidance materials and develop agreements on standards and regulations for observing systems, data, and products. In all cases, adherence to the principles of free and unrestricted exchange of data should be

strongly encouraged, particularly in relation to the designated Essential Climate Variables, which are both currently feasible for global implementation and have a high impact on UNFCCC [UN Framework Convention on Climate Change] requirements.

In the area of nature and biodiversity, the need for improved standards to help production and use of information for EU Directives and international conventions has already been noted. Whilst these need to start at the point of monitoring and survey, they are not confined there – similar standards are required throughout the information chain. Problems in this area are exacerbated by the dispersed nature of the data holdings: for example, species data is typically collected and managed by different thematic groups and, historically, there has been little effort directed towards the introduction of common practices and standards that would achieve greater comparability. Problems also occur in accessing many datasets because of the diversity of sources, poor documentation of delivery routes, inconsistency in content and format, and various legal and institutional constraints on access.

Likewise, issues of data and exchange standards hamper information acquisition and use throughout the emergency preparedness and security domain, for example in relation to humanitarian aid, geophysical hazards, food security and technological hazards. Timeliness of information is crucial in all these areas, as noted before, so any factors that inhibit or delay access to the available information can be serious. The diversity of information needed, and the range of users involved, exacerbates these difficulties, for it implies data exchange between many different organisations. The difficulty in predicting where or when information may be needed adds to the challenges.

LINKS BETWEEN INFORMATION SOURCES, TECHNOLOGIES AND SYSTEMS

Use of extant monitoring and assessment systems

There is a clear need to make best use of the wide range and large volumes of data that already exist. To achieve this requires close links and interchange between extant monitoring and assessment systems, both in Europe and more widely. In general, these links remain to be forged and further developed.

In the case of air quality, the need and opportunity coincide. Procedures for air quality monitoring are already well-established at the national and local level, and various systems are also in place to integrate the information at the European level, in support of the Air Quality Directive and obligations on long-range transport of air pollution. Only a small proportion of the available information currently enters these systems, however, because of the lack of involvement of many of the more local (and some of the national) monitoring networks. To a large degree, this reflects lack of commitment of the EU Member States, and lack of progress at the national level in bringing together information from the various regional or municipal networks. Stronger links between these networks would have multiple benefits, not just to the EU, but also for the local and regional networks themselves, for example by providing wider, contextual information and encouraging the sharing of experience and good practice.

Opportunities for improved linkage between extant monitoring and assessment systems also exist in the areas of urban development and waste. In both of these, there is a plethora of monitoring arrangements in place, especially at the local level (for example, those run by local authorities). Extracting policy-relevant information from these at the national and European level (and, indeed, often at the local level) is made difficult by the poor links between the monitoring systems. Even within the same organisation there is often limited awareness of data held by different departments or appreciation of the benefits to be gained from co-ordinating or exchanging them. The ability to feed this data up to higher levels for policy support thus remains weak. Partly as a result, information on trends and patterns of waste generation, waste treatment, urban land use and urban expansion remains partial, uncertain and poor quality (notwithstanding, for example, the requirements of the Waste Directive).

Integration of Earth observation, ground-based and management-derived data

There are major potential advantages to be gained from linking data from remotely sensed (airborne or space) observing systems with *in situ* monitoring networks, so as to maximise the benefits of the different approaches. Although these technologies are often used independently, in reality they should be regarded as complementary. The problem has been tackled well in operational meteorology through routine data assimilation by numerical weather prediction models and the same approach is being adopted by those seeking to provide operational oceanographic services. The resulting analyses are important resources for policy development, implementation, assessment and research.

Opportunities exist elsewhere to enhance information through linkage and integration of different types of data. Integration of fixed-site monitoring (baseline stations) with more temporary sampling, targeted at specific applications, provides a cost-effective means to expand the scope of existing networks, without the need to invest in new permanent or technically sophisticated monitoring stations. Examples include the use of low-cost passive samplers to supplement air quality monitoring with fixed-site instruments. Similarly, the value of relatively simple survey techniques, targeted at specific issues or areas, can be enhanced by integrating them with major national land surveys. It should also be recognised that a wealth of potentially useful information is collected as part of routine management operations, either for the purposes of organisational monitoring or in accordance with legislation. A lot of health data is provided in this way, for example through routine reporting of patient admissions and discharges by hospitals. Managers responsible for nature conservation and protection typically collect a wide range of information in order to monitor the effects of management, and to help plan and target their activities. Much of the information available on socio-economic activities (for example, agriculture, industrial production, transport) and on emissions to the environment comes from the same sources. Not all (and in some cases very little) of this information tends to be readily available to outside users, though a substantial portion does get aggregated to the regional or national scale, often for the purpose of public policy. Opportunities do exist, however, to collate and tap this information much more effectively.

One area in which this capability for integration is clearly important is oil spill monitoring. Substantial effort is currently going into the detection of oil spills, primarily to enable major oil releases to be combated. Routine monitoring is very patchy, however, with some countries, in the Baltic and Mediterranean for example, providing no information, and some others carrying out surveillance in a manner that does not enable trends to be identified and therefore the success or otherwise of current policy and practice to be established. Therefore, there is a need for, and work is under way to establish, the optimum achievable combination of onshore, ship, aircraft and satellite-based monitoring for operational implementation throughout the EU (DISMAR 2004, Oceanides 2004, Roses 2004).

Similar benefits are likely to accrue in relation to monitoring and modelling of hazards such as wildfires, seismic and volcanic risks, soil erosion and flooding. In each of these cases, there are important, unfilled needs for methods of risk assessment, as a basis both for risk avoidance and for planning emergency response. Satellite-derived information can be vital in this context, for example as a source of data on land cover or to detect scars from past events, but knowledge of the population and asset base at risk is largely unknown. To interpret available information in a prospective way (for example, to quantify the probabilities of future events) historical information is also needed, much of which is likely to be derived from traditional ground-based monitoring (and much, too, from informal records and historic documents). To this also has to be added information on current ground conditions (for example, infiltration capacity in the case of flooding, state of brushwood in the case of wildfires), and associated management activities. In addition, early warning of events requires the further ability to monitor and integrate information on changes on the ground, for example, environmental signals and precursors such as small shock-waves in the case of seismic risks, or soil moisture conditions in the case of floods. Linkage of these different information sources greatly enhances the quality of the resulting risk assessment, and also provides the basis for modelling potential impacts or undertaking what-if scenarios. Such links are not generally in place at present.

Urban development, technological hazards, waste and noise are also areas in which improved linkage between different monitoring systems and technologies is both necessary and achievable. Common elements to all these topics are their localised extent and the limitations inherent in using more traditional ground-based monitoring, for example because of the local diversity of conditions and the costs of monitoring. To some extent, these gaps can be bridged by the use of remotely-sensed data, though important limitations result from the relatively coarse spatial resolution of the data and the fact that many of the features and processes of interest are not directly observable from remote sensing platforms. Management-derived information provides a further important and often relatively under-utilised source of information, for example on waste treatment, emissions and development plans.

Inter-comparison and inter-calibration

It is important to know the extent to which the differences reported by different models or monitoring systems are a genuine reflection of environmental conditions, and not an artefact of the methodology. Inter-comparisons between different methods, sensitivity analyses to assess their robustness to data inputs and other operational factors, validation studies and evaluation and reporting of errors are consequently all important. In particular, there is a need to improve understanding and quantification of the ways in which uncertainties propagate as data pass down the information chain.

In some areas, procedures for inter-comparison and inter-calibration are already relatively well-established. Two good examples are air quality and water quality, where considerable effort has been devoted to comparing the performance of different monitoring systems and transport or dispersion models. Despite this, marked discrepancies still exist between many national and local monitoring networks, and the availability of both monitored and modelled data at the EU level is limited. This is partly because the implications of these inter-comparisons are often not followed up, that is, the relevant changes to monitoring and modelling systems are not adopted at the national or local level. One reason for this is the concern of operators and managers to retain backwards compatibility with earlier records. Marked differences thus persist in the siting of both air quality and water monitoring stations in different countries; the pollutant species and determinants vary substantially and measurement methods are not always consistent or comparable.

INTEGRATION ACROSS POLICY AREAS

Linkage between sectors and policy themes

Linkage across sectors and policy themes faces deep-seated problems, largely reflecting the different approaches to environmental information and monitoring in different sectors and thematic areas. These differences make it difficult to share information or agree on how it should be interpreted. They also make it difficult, as a consequence, to provide a balanced and holistic analysis of many policy issues. Several areas can be recognised, however, within which the need for such analysis is urgent, and in which relevant monitoring systems are largely in place even though the linkage is not. These include linkage of environmental information to:

- industrial activity, for example, manufacturing;

- the operation of the energy industry, including methane leak control in the gas systems that provide the largest source of energy in many European Union accession countries and candidate nations;

- land use and land management, for example, agricultural and forest statistics, chemical and energy inputs;

- demography and social conditions, for example, population numbers, deprivation;

- health, for example, mortality, morbidity; and

- economic conditions, for example, employment, costs.

These are especially significant because, between them, they represent some of the main pressures on the environment.

One area within which these pressures are particularly important is nature and biodiversity. Biodiversity is especially sensitive to land use and land management practices, and these in turn are responsive both to global economic and to EU political influences. However, information on pressures on habitats and species remains weak, and models of how these pressures feed through to affect biodiversity (and how they might respond to changes in policy) are often missing or relatively crude. One major problem in this context is the poor spatial match between socio-economic and land use data on the one hand, and biological recording on the other.

A similar need emerges in relation to hazards such as flooding. Land use and land management within the catchment (for example, afforestation, urban development, soil compaction) directly influence the likelihood and magnitude of floods, and the attendant dangers of soil erosion and sediment transport. All these are also responsive to policy measures in different sectoral areas, for example in agriculture, regional development and forestry. The human, social and economic impacts of floods may also be substantial (for example, through loss of life, destruction of housing, loss of livelihoods and economic costs of compensation, reconstruction and recovery). For all these reasons, there is a need for close integration of information on population, health, social conditions, economic activities, land management, settlements and buildings and flood protection measures, in order to help predict floods and flood risks, assess susceptibility, and to assess and monitor their impacts. Currently, linkage of these different sources of information is inhibited by a number of factors, including differences in the systems of georeferencing used, issues of confidentiality (for example, in relation to land management and economic data) and inconsistencies in definitions.

Exactly the same argument can be made in relation to soil protection. Land use and management practices, not only in agriculture but also forestry, mineral extraction, construction, waste management, industry and transport, all have significant impacts on soil conditions, either directly (for example, by soil stripping, sealing, soil modification) or indirectly (for example, via air pollution). Policies in all of these areas, especially in relation to agriculture and forestry, are clearly major drivers of change. The human and economic impacts of soil damage and loss are also substantial, and in many cases complex. Soil erosion, for example, may not only result in yield reduction and land abandonment or agricultural extensification, but greatly adds to costs of water supply management. Again, linkage and integration of all the different types of information required to monitor and assess these problems are currently difficult, because of the cross-cutting and often unco-ordinated responsibilities of the various agencies involved, and their differing strategies of information gathering and reporting. Closer collaboration between these agencies is a high priority, especially through joint planning of monitoring programmes and improved information exchange.

Linkage of environmental and non-environmental data across different sectors and policy areas is also significant in the areas of noise and waste. In the former, there is a need for more holistic approaches to environmental and health impact assessment that see noise as part of a wider set of risk factors associated with transport and industrial activities. One major need in this context is to be able to integrate information on transport with environmental information in order to develop more reliable models of noise emission and propagation. Links must also be established to information on population, the built environment and human health, in order to assess exposures and health impacts.

Case study: land cover data integration

Human-induced changes in the Earth's surface conditions have affected natural resource availability, biodiversity, atmospheric composition and climate on all geographic scales. Sustainable management of forests, and of other land cover types subject to regimes of rapid change, has emerged as one of the most difficult, serious, and pressing environmental issues.

Land cover studies are cross-cutting and multi-disciplinary. Therefore, it is necessary to maintain close liaisons with global (the carbon cycle) and local (land cover) communities as well as pertinent international organisations. Biomass estimation is necessary in order to extend the land cover and land cover change data to be able to quantify carbon stock and stock changes. Land use as inferred from land cover determines ecosystem impacts. Similarly, quantifying land use intensity (for example, how much land is actually being cultivated) is a key for models of drivers of future land cover and ecosystem change. For validation, close links to federal agricultural agencies need to be structured.

Monitoring of Land Cover Change (LCC) is the next critical element: proof of concept for LCC datasets is already in place (Agbu and James 1994, Ehrlich *et al* 1994, Smith *et al* 1997). The work now needs to be extended to fine resolution data. Especially in areas of rapid change, fine resolution Earth observation products have to be connected to *in situ* observatories, for example forest inventory and assessment data.

Figure 8.1 overleaf illustrates classification discrepancies and the extent of multi-scale land cover variations for a taiga-steppe transition region in southern central Siberia. The left hand map is based on SPOT VEGETATION data and the Global 2000 Landcover product; the middle map is based on Moderate Resolution Imaging Spectroradiometer (MODIS) data and the LC product; and the right hand map is based on Medium Resolution Imaging Spectroradiometer (MERIS) data but also uses the MODIS LC classification system. Each approach shows a different result and can lead to different interpretations of land cover status and LCC.

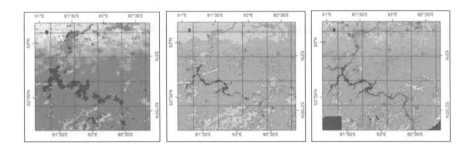

Figure 8.1: Variations in land cover classification depending on the datasets used and the classification scheme

DATA ARCHIVES

The need for archives

An important justification for the creation and maintenance of long-term data archives is the need for reliable historical information on key environmental variables to serve as baselines against which to assess the current situation and as a means of detecting and evaluating significant change and long-term trends. This is a requirement that extends across almost all environmental monitoring themes, but has been singled out as being particularly significant in relation to air quality, flood risk, seismic hazard, noise, soil protection, food security, the provision of humanitarian aid and responses to technological hazards. In the context of environmental and humanitarian disasters, it is extremely important to have available information that can accurately describe conditions pertaining to environment, social and human health prior to the disaster.

The use of reference baselines for the detection and estimation of change has implications for the collection of the datasets used to define the baseline conditions. These implications relate to the consistency and stability of the observing methods and analysis techniques, the estimation of quality and uncertainty, and the continuity of the long-term record.

Case study: the 'repeated baselines' problem

Inventories of natural resources, such as land cover, are expensive and complex undertakings. Many social and political concerns are expressed through policies that are replaced, revised or updated at frequent intervals. When an inventory is commissioned, the user wants something that meets their current needs arising from the new policy, the scientist wants to show their improved understanding of the phenomena, and the technologist wants to use the best and newest technology. Each inventory therefore becomes a new 'baseline' against which change could be measured, but it never is, because next time there will be different policy questions, better science and new technology.

Land cover mapping in the UK illustrates the problem well. The 1930 land utilisation survey, the 1990 national land cover map and the 2000 Co-ordination of Information on the Environment (CORINE) land cover map all purport to be fundamental baselines of land cover in the UK for their respective years. However, each one uses a different framework and spatial scale, and hence they are not easily comparable with each other, never mind as a basis for comparison with other land cover data. While the three maps were individually regarded as baselines in their time, they are not readily comparable and so cannot easily be used for environmental change analysis.

Many examples of this problem can be cited at both European and national levels, in many different areas of analysis, though the problems are often greatest in areas where the technology is changing rapidly (for example, Earth observation). In the UK, for example, several full or partial surveys of land cover have been carried out over the last 30–40 years, using either airborne or spaceborne technologies. Because of changes in classification system, spatial resolution and geographical representation, direct comparisons between most of these are not feasible, and changes in land cover can rarely be measured accurately. Even the last two surveys, using satellite data from the late 1980s and late 1990s, cannot be directly compared. Similar experiences have occurred in other European Member States.

Problems of multiple baselines also potentially occur in other areas where, as yet, only one-off studies, using non-standard methods, have been undertaken, for example the CORINE maps of soil erosion in southern European states. The solution lies in trying to build in backwards compatibility to monitoring and survey systems. Though this may add to the costs of individual surveys (for example, by requiring some duplication of methodology), it can greatly add to the utility of the information, by providing comparability in the long term.

One example of the problem – and how it can be solved – is provided by the BIOPRESS study (BIOPRESS 2004). This is producing a useful, consistent and coherent estimate of LCC across Europe over the last 50 years. It implies the use of an established ontology (in this case, CORINE), consistent input data (air photographs), incorporating uncertainty estimation in the production methodology, and having as its primary objective the identification of change rather than the production of a new inventory.

Environmental indicators

Good quality, historical data is also needed to aid the development of generally applicable and sensitive environmental indicators, and to assess how such indicators respond to actual conditions. Such indicators should be interpreted in the context of a reliable historical record. Any study of European Environment Agency (EEA) environmental assessments rapidly reveals the great difficulty of gaining access to long running datasets of the required quality in almost every field of study, despite the considerable efforts that are expended in the Environment Information and Observation Network (EIONET) to achieve this goal. There are many reasons for this, but the low priority given by those funding research and operational data gathering to the maintenance of long-term datasets to serve future, currently undefined, enquiries is the source of many of them. Without our more enlightened predecessors, there would be a much reduced climate record today, but this example of good practice is largely ignored.

This is not to deny that there are many valuable environmental data archives in Europe. In many cases, however, inputs to these occur on an *ad hoc* or voluntary basis, with the result that data capture is variable and often inadequate for policy-related needs. There is also much to be done to enable existing databases to be accessed in near-real time (for modelling purposes) and in delayed mode. The systems for this are generally not in place. For example, the Global Monitoring for Environment and Security (GMES) Marine Forum in Athens in December 2002 noted that:

> … marine biogeochemical data are highly dispersed and generally without quality control. Concerted actions and networking will be necessary to ensure the widest and fastest access to comprehensive, coherent and compatible datasets by the operational and scientific communities. Only a professional, semi-distributed, multidisciplinary data management infrastructure will be able to give an appropriate access to national data holdings, merge them with new data collected in real time and delayed mode and prepare the best timely integrated data products, that scientific, technical and economic studies require.

Obtaining long-term air pollution data (for example, to assess potential ecological and health impacts with long latency times) faces similar problems. Past data records have rarely been collated in a coherent form, and invariably remain in a non-digital form. Progressive deterioration of many of these records is inevitable as a result of decomposition and degradation of the storage media and loss of institutional knowledge about the datasets. Better stewardship of these records, and efforts to recapture the more important data in a more durable form, are important priorities.

Metadata

Inadequacies have been noted across all environmental monitoring themes in the standards used to record metadata, in the availability of data catalogues and other data discovery tools and in the existence, level of detail and consistency of the documentation needed for effective retrieval, evaluation and use of monitoring data.

Numerous examples can be cited of earlier initiatives in this area that have been supported by the European Commission. These include the CORINE Catalogue of Data Sources (EEA 1999a) and projects such as the European Wide Service Exchange (EWSE), undertaken in the context of the Centre for Earth Observation (CEO) project. Tangible benefits from these projects have still to be demonstrated and a real user community has yet to emerge. Moreover, there is now a plethora of inventories, directories and catalogues of data holdings in the field of environment and, to a lesser extent, security. With the growth of the Internet, this resource is expanding uncontrollably – so much so that an argument could now be made for establishing a catalogue of catalogues. While it is clearly helpful to be able to access rapidly the identifying characteristics of those datasets that are central to the purposes of global environmental monitoring, the argument for yet another open-ended exercise in data cataloguing is by no means proven.

Nevertheless, there is a clear, unfilled requirement for the development and implementation of agreed standards for describing the characteristics of monitoring data. Such metadata standards are important as a tool to ensure consistency and compatibility, not just of the metadata, but also of the actual data. They would also greatly simplify access to and use of both data and metadata.

CONCLUSIONS

Most environmental applications are custom-built, using bespoke data processing elements and unique sensor and computational infrastructures. Because little or no effort goes into making the individual functional components universally accessible, stand-alone custom development tends to be perpetuated. In the long term this approach will waste resources and inhibit fast and robust deployment of environmental monitoring systems.

It is important therefore not to think of global monitoring systems as a collection of stand-alone applications, but as an infrastructure that actively supports the interconnection of components that are already available and promotes the intelligent implementation of new components that can be used by many. It is equally important to realise that this interconnection must take place not only at the physical level (bytes) but also at the semantic level (information and services).

Unlike inconsistencies in survey and measurement methods, differences in practices for data storage, analysis and exchange are to some extent amenable to correction after the event. However, this is often costly and time-consuming, and is in any case only possible if reliable information on data genealogy or pedigree is available. In many cases, such differences are major deterrents to the wider use of data: at best, they slow down access to information and increase its cost; at worst, where timeliness of information supply is crucial, they may compromise effective action.

Examples of the enhancement of information through linkage and integration of different types of data include the following:

- The linkage of remotely sensed (airborne or space) observing systems with *in situ* monitoring networks so as to maximise the benefits of the different approaches. The challenge has been tackled well in operational meteorology through routine data assimilation by numerical weather prediction models, and the same approach is being applied to the provision of operational oceanographic services.

- The integration of fixed-site monitoring (baseline stations) with more temporary sampling, targeted at specific applications, in order to expand the scope of existing networks.

- Integrating into the environmental monitoring process information collected as part of routine management operations, to improve the management of drivers and impacts of environmental change.

It is important to know the extent to which the differences reported by different models or monitoring systems genuinely reflect environmental conditions, and are not an artefact of the methodology. In some areas, procedures for inter-comparison and inter-calibration are already relatively well-established. Despite this, marked discrepancies still exist between many national and local monitoring networks, and the availability of both monitored and modelled data at the EU level is limited, partly through a failure to follow up the implications of inter-comparisons.

It is possible to identify priority areas in which relevant monitoring systems are largely in place even though the linkage is not. These include linkage of environmental information to levels of activity in the industrial and energy sectors, land use and land management, demography and social conditions, health and economic conditions. These are especially significant because they represent between them some of the main environmental pressures as well as some of the key motives for environmental policy.

More generally, linkage and integration of information across sectors is inhibited because of the cross-cutting and often conflicting responsibilities of the agencies involved, and their differing strategies of information gathering and reporting. There is also pressure, particularly from funding agencies and policy makers, for better vertical integration between EU (or global) policies and strategies, national policies, and local programmes and plans. However, there are important unsolved technical issues and questions of subsidiarity and autonomy that need to be addressed.

Good quality historical data is crucial as a baseline against which to assess current states and to detect and evaluate long-term trends. Their value depends on the consistency and stability of the observing methods and analysis techniques, the continuity of the long-term record and the degree to which it is possible to make objective assessments of their accuracy. There is evidence of a paucity of long running datasets of the required quality in almost every sector of environmental monitoring. In many cases where long-term records exist, contributions to these occur on an *ad hoc* or voluntary basis, with the result that data capture is variable and often inadequate for policy-related needs. The low funding priority given to the maintenance of long-term datasets to serve future, but currently undefined, enquiries is a major causal factor.

CHAPTER 9

A EUROPEAN SHARED INFORMATION SERVICE

EUROPEAN SPATIAL DATA INFRASTRUCTURE

As discussed earlier in this book (see Chapter 6), several countries have developed their own spatial data infrastructures, most notably Australia, New Zealand, Canada and the US. Within the context of Global Monitoring for Environment and Security (GMES) there is an interesting question of whether a European Spatial Data Infrastructure (ESDI) can assist in the process of turning environmental data into information. There are clear needs for data and information and there are many environmental information services in Europe, for example the topic centres of the European Environment Agency (EEA) and the Satellite Application Facilities (SAFs) of EUMETSAT. However, while there are some elements in place, GMES has also considered the issues of a European Shared Information Service (ESIS) and the benefits that can flow from improved and co-ordinated access to information.

The concept of an ESIS was discussed at the GMES Forum 3 held in Athens in June 2003. The concept is shown in Figure 9.1 overleaf. The main aims of the concept of an ESIS are to allow users to have better access to good quality environmental data and information, and thereby to contribute to more cost-efficient information services and thus to economic growth. The main functions of an ESIS concept are:

- operational observations;

- operational production and dissemination of information; and

- cataloguing and library access to information.

In order to provide cost-efficient information, and better access to that information, a European shared information capacity will have to consider the challenges of access to data considered in this book. This chapter provides an examination of the data policy issues raised in Figure 9.1. Figure 9.2 is the same concept with an annotation of seven data policy issues that are considered in this chapter. The following sections consider each of the seven annotated parts of Figure 9.2 in turn, namely licences, price policy, encryption, portals, archives, calibration and validation, and standards.

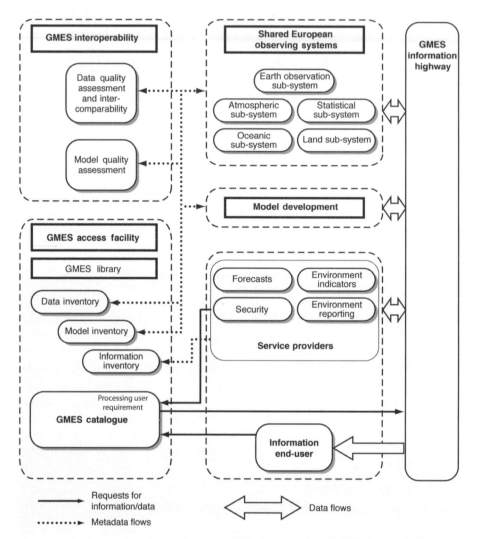

Figure 9.1: The main elements of an ESIS discussed at GMES Forum 3, Athens, June 2003. Source: EC (2003c)

A: Licences

The shared European observing system of GMES envisages data from a wide variety of sources: oceanography, meteorology, terrestrial observations, Earth observation, and statistical agencies, among others. Accessible information on the ownership and licence conditions of the datasets is essential and must be explicit. An ESIS can add value by co-ordinating the arrangements that data suppliers make for licensing and intellectual property rights (IPR) protection for the data

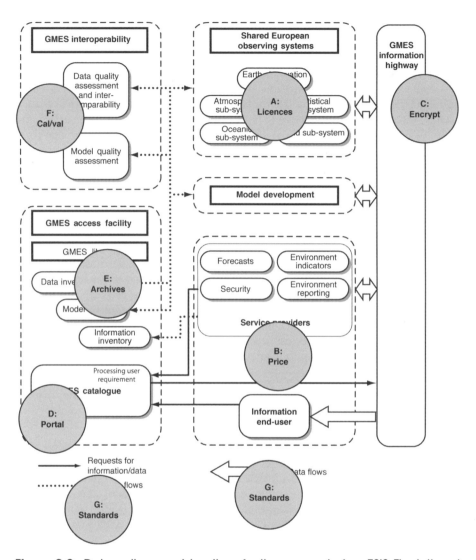

Figure 9.2: Data policy considerations for the concept of an ESIS. The lettered circles link with the text in this chapter

used within GMES. Several GMES Thematic Projects have suggested that GMES could agree improved licence arrangements with data suppliers to allow multiple use of the data held within a European shared information capacity. Some progress has already been made in such a form of 'gateway licence' at the European Commission Joint Research Centre (JRC), which would allow all JRC users access to datasets purchased through the gateway licence.

B: Price policy

The interface with the user is also an interface on price policy, that is, what fees (if any) the user will be charged for the data provided via an ESIS. The funding of operational systems is a key component of global environmental monitoring and was a recurrent theme of discussion during the GMES Forum held in Brussels in July 2002. Pricing policy is central to the funding of operational systems because GMES systems and services have to have sustainable funding in order to become truly operational. Broadly speaking, three scenarios can be advanced that encompass distinct approaches to price policy and hence to operational funding:

- *Scenario 1.* European funding for global environmental monitoring considered as a public good. Environmental data and information could be provided free of charge (or perhaps at marginal cost prices) to users as a raw material. This scenario is used in the World Meteorological Organization (WMO) Resolution 40 and in the data policy of the Intergovernmental Oceanographic Commission (IOC), where essential data for mainstream applications and for research are made available at a marginal cost price. It is also used in the US for all federally-produced data. An interesting but open question in Europe concerns who is to fund the environmental monitoring required to support climate monitoring to fulfil obligations that arise from the findings of the Intergovernmental Panel on Climate Change (IPCC). Such monitoring is seen by many as a government responsibility where the climate monitoring data is provided to society as a public good.

  The European Space Agency (ESA) Oxygen project (O_2 = open and operational) can be noted in this context. The ESA document on the Oxygen project declares clearly that 'inexpensive or even free data is a necessary but not sufficient condition for the development of a mature industry' (ESA 2003).

- *Scenario 2.* A European public sector organisation funds environmental data collection as a customer for its own purposes, and therefore purchases a private good. The European public sector organisation then makes the data widely available. It is possible to envisage government environmental departments acting in this way, that is, funding data acquisition in support of (say) climate change policy and then making the data widely available to other users either for free or at a low cost.

- *Scenario 3.* A multiplicity of end-user customers for regular environmental monitoring. The customers would buy data and information as private goods and the demand would be federated within a European context. A European spatial data service could be the vehicle to provide the data, products and services and so connect the suppliers with the users. Examples of this concept are coastal sea state monitoring and agriculture monitoring at the national level by governments or by private sector organisations.

C: Encryption

The concept of an ESIS as discussed at GMES Forum 3 could exploit the technological possibilities of an information highway. One opportunity here is to provide data to an environmental information highway in an encrypted form so that all users can gain access to the encrypted data. The encryption process can be accompanied by licences to decrypt the data: the decryption keys can be designed in a sophisticated way to be specific on data, products, time periods and spatial locations, and to provide a tailored service to the needs of individual users. EUMETSAT has introduced this type of scheme by broadcasting all Meteosat data, and then allowing open access to data collected at certain times of day while encrypting other time period transmissions. A similar system can be found in the 'pay-per-view' broadcasts of the satellite television operators (see also the discussion of DVB technology in Chapter 7).

Encryption offers an opportunity for efficient distribution of regular, operational data, as is the case with Meteosat data. In the case of data held in archives, a catalogue search is required to find the right dataset, followed by transmission of the data. In this case, the catalogue search could be free of charge and the data transmission achieved via an environmental information highway in an encrypted form.

D: Portals

Catalogues of global environmental data and information have the opportunity to be powerful portals to a wide variety of data and information, and so allow interoperability across different systems and datasets. What should be avoided is users having to examine many data catalogues depending on the dataset and the models they are using. This means that there is a requirement for an efficient interface between an access facility and any shared European observing service.

The experience with portals in the environmental sector has so far been highly varied: some are excellent, particularly those in the Geographic Information System (GIS) sector, and some are poor, including some in the Earth observation sector. The Geospatial One-Stop portal (FirstGov 2004) in the US is an excellent example of a successful data portal, and was created to provide data for the US Spatial Data Infrastructure. Key design goals of the portal were to allow users to share and find data and services with minimal effort, plus enhanced search capabilities to allow users to track down datasets. The portal provides users with data for use in GIS software, geospatial applications and information on events and activities. Data categories covered by the portal include atmosphere and climate, human health and disease, and utilities and communication.

E: Archives

Access to past environmental data is often poor. Historical environmental data is often not in digital form. There are physical problems with both digital tapes and tape reading machines because of obsolescence and age. However, there is great value in safeguarding environmental data over the long term, including the following:

- to reconstruct the past;

- to obtain insight into and understanding of the past;

- to develop models of environmental processes; and

- to predict the future.

Improved access to high quality archives can increase the value of contemporary data, for example in conflict avoidance and humanitarian aid applications. A challenge for global environmental monitoring is that environmental data archives are not cost-effective when seen in a short-term, micro-economic view. A long-term perspective is essential, supported by funding. In addition, there must be efforts to secure environmental datasets that might otherwise be lost.

F: Calibration and validation

One element of global environmental monitoring is modelling, an important part of the process of converting data into information and information into knowledge. For model output to be valid and useful there is a requirement for data calibration and validation of results. The experience of methane monitoring in Europe shows the importance of high quality calibration and validation to provide data that is reliable. An ESIS will need to have calibration and validation structures in place to enable data to be turned into information.

G: Standards

A more coherent structure of standards is necessary for enabling efficient data flows and metadata flows. There is a need for agreement on:

- product standards, for example land cover types or marine oil slick identification;

- standards for data flows and for files on an environmental information highway; and

- implementation of metadata readers and interpreters.

Market standards can make a positive contribution to mature global environmental monitoring systems. For data and information to be most accessible and useful to users, data should be provided in the most widely-used formats. A good example can be seen in the GIS sector where industry-standard formats associated with widely-used packages such as ArcGIS are frequently used. Similarly, meta-languages such as Extensible Markup Language (XML) are being widely used on the web. XML has been rapidly adopted by science, industry and e-commerce and has already spawned new metadata standards in fields such as mathematics, chemistry, astronomy, multi-media and web micro-payments (Houlding 2001). Using the most widely-accepted market standards is likely to encourage greater use of environmental services, as data will be provided in formats that are familiar and widely-accepted by users.

PEOPLE

The discussion in this chapter has been about legal and technical policy issues. This reflects the concepts for sharing information as an information system or infrastructure. If data and information sharing were to stay at the level of a data system, or a network of data systems, then its impact would be limited. Data systems are powerful, effective and efficient when they are designed for a specific purpose to meet a specific need: computer systems in banks, for example. Where they are designed with general purpose information in mind, then the challenge is greater and more complex, and their effectiveness much weaker.

The past emphasis on computer and infrastructure systems neglects the importance of people. For successful global environmental monitoring, it is essential that people are involved as a major part of systems or services. Any investments in new European systems and in connecting existing data systems would provide the technology foundation on which people will work. People are needed to assist users with finding data, with understanding catalogues, with interpreting models, with formatting datasets, with checking quality, with establishing licence agreements, with investigating existing archives and with building coherent archives. In Europe, people are vital for communicating in different languages in different parts of Europe. All environmental monitoring systems must put people at their heart.

INSPIRE

The initiative

During 2002 and 2003 an initiative on Infrastructure for Spatial Information in Europe (INSPIRE) was launched by the European Commission and developed in collaboration with the Member States and accession countries. INSPIRE aims at making available relevant, harmonised and high quality geographic information to support the formulation, implementation, monitoring and evaluation of Community policies with a territorial dimension or impact to ensure a better knowledge-based European environment policy (Perera 2003).

The European Commission is preparing a proposal for a Directive of the European Parliament and of the Council aiming at the establishment of an ESDI. The elements of this ESDI consist of: the spatial data; the metadata; spatial data services; technologies, guidelines and specifications; agreements on sharing, access and use; organisations; monitoring mechanisms; processes and procedures; as well as the human and financial resources related to the European initiative. The European legislation will mainly concern those spatial data infrastructure elements which fall under the jurisdiction of EU Member States' territories or Exclusive Economic Area, and which are held in electronic format by or for public authorities. Approximately a dozen priority spatial data themes will be subject to a full range of requirements for harmonised specifications (for example, co-ordinated reference systems, transport and hydrology networks, elevation). Other priority themes (for example, soil, geology, natural risks) will be concerned mainly with the standardisation of their spatial features but not with their thematic content. Figure 9.3 shows a summary of the INSPIRE structure as regards information flow from the data resources of the ESDI, via the INSPIRE key characteristics, to the broad-based user community.

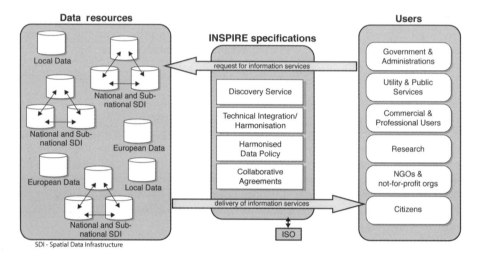

Figure 9.3: The INSPIRE structure and information flow. Source: INSPIRE (2004)

In early 2004 the European Commission developed a revised version of INSPIRE (EC 2004b). Two task forces were established: one led by the EEA was charged with assessing the scope of INSPIRE, while a second task force was led by Eurostat to assess the impacts of INSPIRE (EC 2004b). The primary focus of INSPIRE has become the basic data necessary to support European Commission sectoral and structural policies, set in the context of a European spatial data infrastructure. This is similar to the concept of fundamental spatial data in Australia and New Zealand (IDC 2001). The basic datasets are likely to comprise administrative units, transport networks, hydrography, elevation, cadastral parcels, land cover, protected sites and ortho-imagery.

INSPIRE and data policy

An important part of the INSPIRE initiative is concerned with a harmonised data policy. In the development of INSPIRE during the year 2002, a working group on data policy and legal issues was established. The report from this working group (EC 2002) provided one of the foundation stones for the subsequent phase of INSPIRE. The working group report recommended the following five basic principles that should be used to improve access to spatial information in Europe:

(1) Data should be collected once and maintained at the level where this can be done most effectively.

(2) It must be possible to seamlessly combine spatial information from different sources across the EU and share it between many users and applications.

(3) It must be possible for spatial data collected at one level of government to be shared between all levels of government.

(4) Spatial data needed for good governance should be available on conditions that do not restrict its extensive use.

(5) It should be easy to discover which spatial data is available, to evaluate its fitness for purpose and to know which conditions apply to its use.

INSPIRE consultation process

During 2003 the INSPIRE programme undertook a consultation process. The paper that formed the key document in the consultation process (EC 2003a) identified a set of 14 policy measures that were proposed for the INSPIRE legislation. The results of this consultation process show a concern by many of the respondents with issues of data policy. The list below summarises from the consultation report the data policy issues that are directly relevant to global monitoring and to an ESIS:

• Almost all respondents (97%) agreed with the five basic INSPIRE principles (see above).

• A majority of respondents (79%) believed that the general interest in the creation of an infrastructure for spatial information justifies funding by public authorities, and mainly from the EU.

• A large proportion of respondents (85%) agreed with the need to establish a common data policy framework to share spatial datasets between public bodies.

• The majority of participants (82%) agreed on the need to establish a general licensing framework for spatial data that goes beyond the public sector.

• A large proportion of respondents (77%) thought that additional legal initiatives will be needed in future to ensure full European coverage of datasets.

• Almost all the respondents (95%) agreed that common specifications and the building of bridges between existing datasets and these common datasets are useful for increasing the potential for re-using public sector spatial data.

• Almost all the respondents (96%) agreed that certain information regarding standards and key components of data should be made available free of charge and with no restrictions on use in order to encourage their use by a wide range of data providers.

- Almost all respondents (94%) agreed that the Member States of the EU should set up services that make it possible to publish, discover, view, access and trade spatial datasets in accordance with common standards.

- Most respondents (71%) favoured the development of a single access point or portal for data and services, implemented on top of national access points.

- Almost all respondents (95%) believed it is important to be able to view the available data, and 81% of these respondents thought it should be free of charge for citizens, NGOs and public authorities.

There is much here that can be learned. The most important message is the desire for the public sector to take the lead in providing spatial information for wide use in Europe, and for the public sector to be an important partner in achieving sustainable funding.

CHAPTER 10

INSTITUTIONAL FACTORS

Nina Costa, Oliver Greening and Zofia Stott

INTRODUCTION

A key attribute of Global Monitoring for Environment and Security (GMES) is that it is a European level initiative. This chapter provides a summary of the European organisations that are concerned with data and information supply for environment and security policies at European level and which would thus be involved in the GMES, or at least candidates for involvement. The detailed description of these European organisations serves the following three functions:

(1) to understand how these European organisations operate currently in terms of decision-making, information flows and financial arrangements;

(2) to review whether these arrangements might need to be altered for these organisations to become involved in GMES; and

(3) to provide lessons for the development of GMES.

In order to facilitate the assessment, this chapter considers the key European players in the three main groups: data suppliers, information suppliers, and end-users or policy makers with a focus on the supply side. This represents a principal categorisation since some of these organisations have multiple roles, that is, some organisations are both suppliers and users. The European organisations investigated in this chapter are:

• European Space Agency (ESA);

• European Organisation for the Exploitation of Meteorological Satellites (EUMETSAT);

• European Environment Agency (EEA);

• Eurostat, the Statistical Office of the European Communities;

• European Union Satellite Centre (EUSC);

• European Commission Joint Research Centre (JRC);

• European Centre for Medium-Range Weather Forecasting (ECMWF);

• EuroGeoSurveys, an association of European geological surveys;

- EuroGeographics, the association of European national mapping and cadastral agencies; and

- European-Mediterranean Seismological Centre (EMSC).

For these European organisations, this chapter reviews their objectives, their decision-making flows, their information flows and their funding, as well as what agreements, if any, exist between these organisations regarding co-operation or data sharing, and links to national and international bodies. The intention is to give an overview of the situation in Europe concerning the key institutions involved in the monitoring of environment and security, centred on European level institutions but also taking into account how these organisations interface with organisations at the national and international levels. The chapter finally discusses the potential roles that such organisations could play in GMES, and the likely constraints that they have regarding participation.

EUROPEAN SPACE AGENCY

ESA objectives

ESA is an intergovernmental organisation, whose mission is to shape the development of Europe's space capability and ensure that investment in space continues to deliver benefits to the people of Europe. In particular, its strategic objectives are: the pursuit of scientific knowledge; enhancing the quality of life on Earth; successful European co-operation; and promotion of European industry. It achieves its objectives by providing and promoting, for exclusively peaceful purposes, space science, research and technology, and space applications.

ESA's main Earth observation programme is Envisat, a multi-instrument mission launched in March 2002 at a cost of approximately two billion euros. On top of this, ESA has two major Earth observation programmes that will provide data relevant to environment and security:

(1) The Earth Explorer element, under the Earth observation Envelope Programme, predominantly concerned with Earth system scientific research, but also providing opportunities to develop new technologies.

(2) The Earth Watch programme, concerned with the development of pre-operational satellites (for example, a first of a series). ESA GMES funding is currently directed through the Earth Watch programme.

Figure 10.1 shows the primary institutional links for ESA, focusing on Earth observation.

Decision flows

ESA has 15 full Member States (Austria, Belgium, Denmark, Finland, France, Germany, Ireland, Italy, the Netherlands, Norway, Portugal, Spain, Sweden, Switzerland and the UK), three co-operating members (Canada, Czech Republic and Hungary) and two associate members (Luxembourg and Greece). The

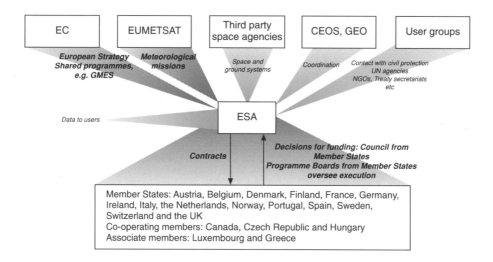

Figure 10.1: Key institutional links for ESA, focusing on Earth observation

rationale for ESA is based on the premise that by co-ordinating the financial, intellectual and industrial resources of its members, ESA can undertake programmes and activities far beyond the scope of any single European country.

ESA is an independent organisation, although it maintains close ties with the European Commission with which it shares a joint space strategy and EUMETSAT for which it develops and builds meteorological satellites. ESA also maintains close links with space organisations outside Europe as part of joint programmes and co-ordination activities (for example, the Committee on Earth Observation Satellites, CEOS, and the Group on Earth Observation, GEO).

ESA's governing body is the Council. This is composed of high-level representatives of ESA Member States and is responsible for drawing up the European Space Plan and ensuring the long-term funding of the Agency's activities. Each ESA Member State has one vote and is represented by a Council delegate from the ministry responsible for space activities in each Member State. In general, Council meetings are held every three months at delegate level and every two to three years at ministerial level. The Council is responsible for drawing up the European Space Plan, ensuring that it is being followed, approving both ongoing and future programmes, and deciding on the level of resources to be made available to ESA.

Oversight of Earth observation activities by Member States is maintained through the Programme Board for Earth Observation (PB-EO). The Programme Board consists of representatives of ESA Member States. The tasks of the Board are threefold:

(1) to monitor and take decisions relating to the execution of Earth observation programmes, except for those decisions that remain within the competence of the Council;

(2) to examine the respective proposals for future programmes and prepare the decisions that remain within the competence of the Council; and

(3) to provide for the co-ordination of European and national Earth observation activities.

Both the PB-EO and the ESA Executive may create advisory bodies, as appropriate. For example, the PB-EO has the Data Operations Scientific and Technical Advisory Group (DOSTAG) as an advisory group. The ESA Executive uses the Earth Science Advisory Committee (ESAC) as a peer review advisory body regarding the choice of future Earth Explorer missions.

ESA's main responsibility is for research and development of space projects. Once declared fully operational and sustainable, these projects have, to date, been handed over to outside bodies for their long-term operational exploitation. In Earth observation the key example is meteorology, where ESA carries out research and development of new systems but operations are financed and carried out by EUMETSAT. The exact responsibilities of each organisation vary from programme to programme, but in general EUMETSAT is in sole charge of ground system activities, whereas responsibility for the space segment gradually transfers from ESA to EUMETSAT as programmes mature. This has implications for national funding. For example, ESA funding is generally derived from national research and development sources, whereas EUMETSAT funding comes through National Meteorological Services (NMSs).

To date, ESA's decision-making process has been very successful in the field of Earth science and technology development satellites (ERS–1, ERS–2, Envisat and now Earth Explorer missions) and meteorology (in collaboration with EUMETSAT), where clear processes have been developed in response to coherent requirements. It is proving more difficult to federate requirements at a European level in the context of new (pre-)operational missions. So today the Earth Watch programme, including the GMES Services Element, is at a preliminary stage and mission activity is coming to the fore at national level in Europe, for example Pleiades in France, Cosmo Skymed in Italy, Rapideye, SAR Lupe and Terra SAR-X in Germany, and the Disaster Monitoring Constellation in the UK.

Information flows

ESA makes data and information products available from its own missions (European Remote Sensing Satellite (ERS), Envisat) and also from certain third party missions. The ESA Earth Observation User Services website (ESA 2004a) provides bibliographic information, catalogues, software tools and product guides which allow users to discover resources of interest.

Acquisition of and access to ESA data and products is controlled through a data policy (ESA 1998, Harris 2002) that defines two categories of use of ESA Earth observation data (see Chapter 4 of this book). Under Category One use, applications are approved by ESA and the users interact directly with ESA for ordering new acquisitions or archive data. This also includes Announcement of Opportunity mechanisms where data is released free of charge. Category Two use applications must interact through commercial distributing entities. In both cases, data acquisition, processing and archiving are carried out in either ESA's own facilities or facilities operating under contract to ESA. While the categorisation is simple, the data policy is complex: it attempts to cater for every eventuality, because data from a wide variety of Earth observation instruments can, in principle, be used in many different ways (near-real time versus off-line, regional versus global) and by many different users (research, operational, commercial).

ESA has numerous data sharing agreements for ERS and Envisat data with Member States (for example, with National Stations for data acquisition) and other countries (for example, Foreign Stations), all defined within the data policy documents.

Today, and perhaps of particular relevance in the GMES context, a number of specialist services are emerging, particularly where near-real time (NRT) access is required. In accordance with the Envisat data policy and the related implementing guidelines, ESA is caring for delivery of near-real time data and certifies the services provided by National Stations also interested in NRT delivery. Two examples are given below:

(1) Near-Real Time Averaged Sea Surface Temperature. This service, operated by the Tromso Ground Station, delivers NRT daily update maps, monthly averaged maps and animations of sea surface temperature from the Along Track Scanning Radiometer (ATSR) instrument free of charge over the web.

(2) Global Ozone Monitoring Experiment (GOME) Fast Delivery Service. The GOME Fast Delivery Service, run by the Dutch meteorological service KNMI, provides NRT ozone column data, profiles and assimilated global ozone maps from GOME data. The KNMI system is able to provide this service within a few hours of data acquisition, meeting the growing demand for ozone products for purposes such as assimilation in numerical weather prediction models, radiation forecasts and ozone measurement experiments. Other examples include fast delivery to meteorological offices for assimilation into weather forecasting models, and collaboration agreements with disaster information and management agencies such as Reuters Foundation AlertNet.

In addition to making its own data available, ESA also acquires, processes and distributes data from third party missions (for example, Landsat, SPOT, SeaWiFS, MODIS, NASA-Aqua, NOAA missions) through its Earthnet programme. A recent significant development in the area was ESA's offer to develop ADEN, a European node for NASDA's ALOS satellite.

ESA's new vision for Earth observation, as outlined in the Oxygen initiative (ESA 2003) is that in the future the data and information systems will be based, as much as possible, on re-use and integration of existing facilities and services, thereby ensuring a maximum sharing of resources and an easy and simple access point for all Earth observation data and all users.

Funding

ESA's total budget for 2002 was €2,853 million, of which €376 million (approximately 13%) was allocated to Earth observation activities. ESA employs approximately 1,800 staff drawn from Member States. About 90% of ESA's budget is spent on external contracts with European industry, and 10% on internal costs.

ESA's funds come from Member States. ESA's mandatory activities (space science programmes and the general budget) are funded by financial contributions from all the Agency's Member States, calculated in accordance with each country's GNP. In addition, ESA conducts a number of optional programmes, including programmes in Earth observation. Each Member State decides in which optional programmes it wishes to participate, and the level of funding it wishes to contribute. Industrial policy is based on *juste retour* principles that ensure that contributions by Member States are balanced by industrial return or contracts to that country. In the case of meteorological satellites, funding also comes from EUMETSAT.

ESA and GMES

ESA is the co-initiator of GMES together with the European Commission. Its Earth Watch programme funds the development of pilot services as part of the GMES Action Plan. ESA will be a key partner in GMES, as it will develop and build new satellites, contribute to the development of Earth observation-based information services, maintain archives which could provide core datasets for the data integration and information management component of GMES, and foster better access to data.

EUMETSAT

EUMETSAT objectives

The role of EUMETSAT is defined by an international convention, which came into force in 1986 and states that the primary objective of EUMETSAT is to establish, maintain and exploit European systems of operational meteorological satellites. A revised convention, which came into force in November 2000, adds a further objective: to contribute to the operational monitoring of the climate and the detection of global climate change. In each case, the operational, long-term nature of the objectives are stressed, giving a clear direction for EUMETSAT's role.

The EUMETSAT convention empowers the EUMETSAT Council to establish satellite programmes meeting these objectives, addressing the requirements of Member States and taking into account, as far as possible, the recommendations of the World Meteorological Organization (WMO).

EUMETSAT can trace its history to the launch of the first Meteosat satellite by ESA in 1977. The success of that satellite led to an intergovernmental conference, held in two sessions in 1981 and 1983, which agreed to establish a new organisation. EUMETSAT took over formal responsibility for the Meteosat system in January 1987. In 1991 it initiated a new programme, Meteosat Second Generation (MSG), to ensure continuity of observations from geostationary orbit until the latter half of the second decade of the 21st century. A Meteosat Transition Programme (MTP) was also commenced, to ensure continuity between the then Meteosat Operational Programme and MSG. This included one further satellite of the same design as its predecessors and a completely new ground system. In 1995, Meteosat operations were transferred to a new ground system and a dedicated Mission Control Centre in EUMETSAT's headquarters in Darmstadt, Germany.

Figure 10.2 shows the primary institutional links for EUMETSAT, balancing its responsibilities for its Member States with those for the international organisations in satellite meteorology.

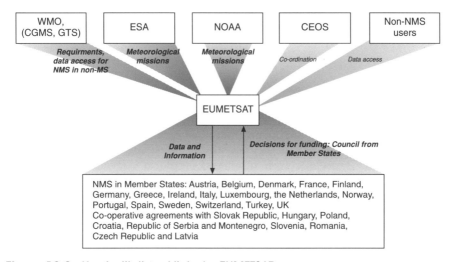

Figure 10.2: Key institutional links for EUMETSAT

Decision flows

EUMETSAT is an intergovernmental organisation of 18 European Member States: Austria, Belgium, Denmark, Finland, France, Germany, Greece, Ireland, Italy, Luxembourg, the Netherlands, Norway, Portugal, Spain, Sweden, Switzerland, Turkey and the UK. These states fund the EUMETSAT programmes and are the principal users of the systems. EUMETSAT also has nine co-operating states: Croatia, Czech Republic, Hungary, Latvia, Poland, Republic of Serbia and Montenegro, Romania, Slovak Republic and Slovenia. The EUMETSAT Council is the supreme decision-making body of the organisation. Each Member State is represented and the national delegation includes participants from the respective NMS. The Council meets at least twice a year and guides the general policy of EUMETSAT and its financial commitments. The Council is supported by several subsidiary delegate bodies, including the Policy Advisory Committee (PAC), the

Scientific and Technical Group (STG), the Administrative and Finance Group (AFG) and the Working Group on Data Policy (WGP). These bodies also meet two to three times a year, preparing the decisions of the full Council. The day-to-day work of EUMETSAT is carried out by the Secretariat located at the EUMETSAT headquarters in Darmstadt, under the control of the Director-General, who is the legal representative of the organisation. Staff are recruited from Member States.

In addition to the meteorological services of Member States and co-operating states, EUMETSAT works in close collaboration with:

- ESA and the National Oceanic and Atmospheric Administration (NOAA) in the development of their satellite programmes;

- WMO and in particular the Co-ordination Group for Meteorological Satellites (CGMS) on co-ordination of requirements for satellite meteorology; and

- the European Commission, on specific projects such as PUMA (Preparation for the Use of MSG in Africa).

Information flows

EUMETSAT's systems are intended primarily for the NMSs of Member States to satisfy their weather forecasting requirements. The NMSs in turn distribute the data to other end-users. The provision of forecasts and satellite images on television several times a day means that most of the citizens of Europe make direct use of EUMETSAT's imagery. Priority is also given to the NMSs of non-Member States. These are given privileged access to data in the continuing tradition of data exchange between meteorological services. They too use the data for the preparation of forecasts and for distribution to television audiences.

As well as these two important categories, there are many other users. Universities and research institutes use EUMETSAT data for research and education. Commercial organisations also use the data, either as end-users (such as airlines) or as service providers (television stations and commercial weather forecasting firms). Smaller Meteosat reception facilities are installed in schools, flying clubs, marinas and by many private individuals. In all, a few thousand systems, located in over 100 countries, are installed for the direct reception of EUMETSAT image data. Of these, over 400 are high performance Primary Data User Stations for which an annual licence fee may be payable to EUMETSAT.

EUMETSAT is working with ESA to create the EUMETSAT Polar System (EPS). This represents the first European owned polar orbiting system of meteorological satellites. The EPS data policy has yet to be fixed, but the system plans to transmit data streams to user stations throughout the world, and by this means users will be able to receive local data in real time from the satellite each time it passes overhead or close to the station. In addition, there will be on-board storage and reception of each orbit at the primary ground station. The data streams will be co-ordinated with those of the NOAA meteorological satellites, but due to the evolving technology and different phasing of the systems the transmission details will differ.

In 2002, EUMETSAT agreed to be an equal partner with Centre National d'Etudes Spatiales (CNES), NASA and NOAA in the Jason-2 Programme for precise ocean surface topography. Unlike Meteosat and EPS, Jason-2 is an optional programme for Member States, with each subscribing at its own selected level.

EUMETSAT has worked with Member States in recent years to develop the Satellite Applications Facilities (SAFs). SAFs make use of Member States' expertise to process data from geostationary and polar orbiting satellites for the generation of meteorological and geophysical products. These are beyond the standard products generated centrally by EUMETSAT. SAF data, products and services are part of the official EUMETSAT ground segment, and will be available on the same basis as the other data, services and products.

At the end of 2003, the following SAFs were under development or ready to commence operations:

- support to nowcasting and very short range forecasting;

- ocean and sea ice monitoring;

- ozone monitoring;

- climate monitoring;

- numerical weather prediction;

- GRAS (GNSS Receiver for Atmospheric Sounding) meteorology; and

- land surface analysis.

Funding

Today EUMETSAT is an organisation of approximately 200 people drawn from its Member States, with an annual budget of around €300 million, with some 90% of that budget being spent on the main programmes. EUMETSAT derives the vast majority of its funding from the contributions of its Member States. These contributions are calculated as pro rata to the Gross National Income (GNI) of the respective states. This justifies the voting structure, which is on the basis of one vote for each country and a weighted majority voting on key financial issues. EUMETSAT operates a best value for money policy with respect to placing industrial contracts for system development and satellite launch, with a stated preference for European technology.

EUMETSAT and GMES

EUMETSAT's data, products and services – both those provided centrally and those available via its network of SAFs – are directly applicable to GMES services. They cover a number of environment and climate topics that are relevant to GMES. EUMETSAT's extended remit, which covers the monitoring of climate change, is of particular relevance to GMES, as currently it is the only organisation that has taken on this role at European level.

EUROPEAN ENVIRONMENT AGENCY

EEA objectives

The EEA is a European Community body with the aim 'to establish a seamless environmental information system' (EEA 1999b). More specifically, the mission of the EEA aims:

> ... to support sustainable development and to help achieve significant and measurable improvement in Europe's environment, through the provision of timely, targeted, relevant and reliable information to policy making agents and the public. (EEA 2002)

The EEA was set up in May 1990 by Council Regulation and commenced operations in 1994. The main objectives of the EEA in carrying out this mandate are twofold:

(1) To assist the Community and member countries to identify, frame, prepare and implement sound and effective environmental policy measures and legislation, and to monitor, evaluate and assess actual and expected progress in the implementation and results of such measures.

(2) To establish and co-ordinate the European environment information and observation network based on an information infrastructure for the collection, analysis, assessment and management of data shared with European Commission services, EEA member countries and international organisations, agreements and conventions.

The Agency's work is in part based on the Earth Information and Observation Network (EIONET), an information network of over 600 environmental bodies and agencies, plus public and private research centres across Europe. It consists of: European Topic Centres (ETCs), covering air and climate change, inland and marine waters, terrestrial environment, nature and biodiversity protection, waste and material flows; National Focal Points; National Reference Centres; and Main Component Elements. Together they are collectors, interpreters and suppliers of environmental data and those with expertise in environmental science, monitoring or modelling.

Decision flows

The EEA has 31 member countries and was the first EU body to open its doors to the 13 countries in central and eastern Europe and the Mediterranean basin that have applied for membership of the EU. Member countries are:

- the 15 European Union Member States in 2003;

- Iceland, Norway and Liechtenstein, which are members of the European Economic Area; and

- Bulgaria, Cyprus, Czech Republic, Estonia, Hungary, Latvia, Lithuania, Malta, Poland, Romania, Slovenia, Slovak Republic and Turkey.

The EEA's products and services are primarily targeted to support policy processes in European Commission services and other EU institutions, in countries and international conventions. The work of the Agency covers the following three main areas:

(1) Networking – stimulate the development and interconnection of Europe-wide environmental data gathering and processing through EIONET, co-operation programmes and international relationships.

(2) Monitoring and reporting – offer reliable, cohesive, simple, low-cost routine monitoring and establish a regular indicator-based reporting system on Europe's environment.

(3) Reference centre – facilitate environmental action by acting as a centre of excellence and clearing house for environmental data and encouraging harmonisation of methods and measurements.

The EEA has an extensive network of collaboration and co-operation with international and European partners, with the aim of improving relevance and efficiency of information, building synergies, supporting countries better and avoiding duplication of activities. The key partnerships are illustrated in Figure 10.3. These are based on Memoranda of Understanding or co-operation frameworks, and the main forms of co-operation are described below.

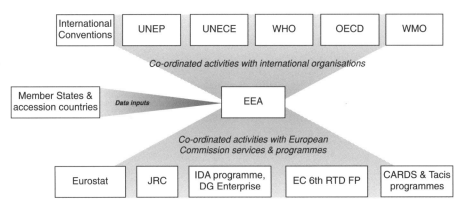

Figure 10.3: Key institutional links for the European Environment Agency

JRC collaboration with EEA's European ETCs is recognised as one of the most important and successful results of recent co-operation. A clear division of activities is leading to mutually beneficial co-operation which streamlines support to Commission services: the JRC focuses on quality control and the EEA on networking with countries, synthesis and assessment. Overall, closer collaboration is being sought to allow a more efficient joint approach to supporting policy and the associated reporting system, particularly in light of the enlargement of the European Union. This includes: more joint work in ETCs; indicators; knowledge management; uncertainty and sensitivity analysis; futures studies; Eurostat's statistical (for example, waste statistics regulation), environmental accounting and indicators projects; and links to the European Commission's sixth framework programme.

Co-operation with the Council of Europe (CoE) on nature conservation, biodiversity and landscape led in 2001 to the signature of a memorandum of understanding between the CoE and the EEA. The co-operation includes work supporting the Bern Convention and its Emerald network, which will complement Natura 2000 in non-EU countries, as well as work on the European Landscape Convention (adopted in 2000) and on the pan-European biological and landscape strategy.

In 2001 a joint meeting between the Organisation for Economic Co-operation and Development (OECD), Eurostat and the EEA was held, to review their complementary data collection activities. This was part of the continued efforts to streamline data flows and reduce the burden on countries to deliver data. The EEA continued to participate in the regular OECD Working Group on Environmental Information and Outlooks.

The co-operation between the EEA and the countries of Eastern Europe, Caucasus and Central Asia (EECCA) takes place in the context of the Environment for Europe process whose secretariat is held by UN Economic Commission for Europe. Acting on behalf of the European Commission, the EEA has the responsibility of producing pan-European reports on the state of the environment, the most recent one being presented at a ministerial conference in Kiev in 2003. In addition, the EEA provides some support for strengthening and streamlining existing capacities in order to improve the reliability and consistency of environmental monitoring, reporting and assessment in the countries and at the pan-European level; it also seeks to enhance public access to environmental information in these countries, as well as delivering environmental information required for international activities.

Regular reporting on the state, trends and outlook of Europe's environment is a common task of the United Nations Environment Programme (UNEP) and the EEA. The long-established co-operation aims to streamline these activities, provide mutual support and avoid duplication of activities. With the enlargement of the Agency, this co-operation is taking on increasing importance, in particular regarding the European region and the Mediterranean region.

The framework of co-operation between the World Health Organization (WHO) and the EEA is defined by common information needs with the aim of introducing health more systematically into environment. This is reflected in the developing co-operation on indicators, which is aimed, in particular, at establishing common indicators in environment and health for regular reporting.

Information flows

Several types of products and other outputs have been developed for the general public and are made available through the EEA website (EEA 2004):

- Reports and publications about Europe's environment and the activities of the Agency.

- Indicators that measure developments in selected issues, including progress towards agreed environmental targets. A core set of indicators is under discussion with EEA's member countries.

- A data service which provides access to datasets used in EEA reports, typically aggregated to country level. This is currently an experimental service providing access to a limited number of datasets under a set of terms of use.

- Reportnet, the EEA's framework for environmental data exchange. This project follows the Directorate General (DG) of the Environment's initiative on review of reporting aiming at streamlining data flows according to the principle: *collect once, use many*. Reportnet functions are underpinned by several information technology tools. The content is built along EEA's priority data flow areas. 170 reporting activities from member countries are presently covered and Reportnet will be extended on the basis of the outcome of the core set of indicator discussions. The Reporting Obligations Database (ROD), which covers both legal and moral obligations, is part of Reportnet.

- Sustainability Targets and Reference (STAR) database, an inventory of current environmental policy targets and sustainability reference values.

- Database on economic instruments used in environmental policy.

Funding

The EEA has an annual budget of up to €30 million which supports around 100 staff and five ETCs. Of the total budget of €30 million, €21 million is drawn from the European Commission and €7.3 million from new Member States. Small amounts flow from the European Free Trade Association (EFTA) and from other sources (EEA 2003).

The EEA and GMES

The EEA has considerable experience in producing information on the state of the environment for policy makers and has already set up an extensive information system comprising 600 nodes across Europe (EIONET). The EEA favours an open access data policy. For example, the EEA has decided (with the European Commission, the JRC and member countries) to make its CORINE Land Cover (CLC) 2000 dataset and IMAGE 2000 an ortho-rectified mosaic of satellite images across Europe, freely available for all non-commercial uses.

The EEA is currently involved in three key areas that have a strong territorial dimension – biodiversity, water and land (land use and land cover changes). The Agency's plan within the 2004–08 strategy is to establish a shared, integrated spatial information system (see also Chapter 9 of this book) in line with the Infrastructure for Spatial Information in Europe (INSPIRE) and GMES initiatives and the need to strengthen support to environment and related policies through the provision of spatial analysis. The central concept will be that of an integrated platform for land use, biodiversity and water, based on the development of Geographic Information Systems (GISs), and the application of environmental accounts and available modelling methodologies.

EUROSTAT

Objectives

As one of the DGs of the European Commission, Eurostat's mission is to provide the EU with a high quality statistical information service, which means statistics that are impartial, reliable and comparable between Member States. Eurostat's key role is to supply statistics to other DGs and hence supply the Commission and other European institutions with data so that they can define, implement and analyse Community policies. Data collection is not carried out by Eurostat itself, but by the statistical authorities in Member States. The Statistical Law of 1997 defines the division of responsibility between national and Community statistical authorities. Eurostat's role is to develop and agree methodologies, conventions and standardised definitions to ensure data comparability, and then to consolidate European statistics. Eurostat developed the European Statistical System (ESS) as a network to lead the way in harmonisation of statistics in close co-operation with national statistical authorities and international organisations. The main institutional links for Eurostat are shown in Figure 10.4.

In the area of the environment, Eurostat has been co-operating with the OECD since 1988 in the treatment, analysis and publication of data collected by both organisations via their biennial joint questionnaire on the state of the environment, covering the topics of waste (generation, treatment), water (resources, abstraction, use and treatment), environmental expenditure and land

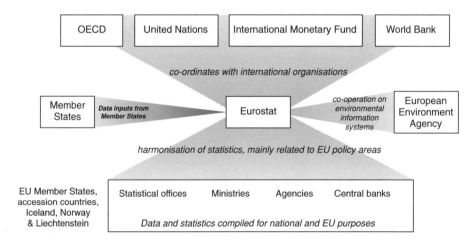

Figure 10.4: Key institutional links for Eurostat

use. Eurostat has been using data from this questionnaire since 1991 for EU Member States and accession states. Other environmental topics of interest to Eurostat include the development of indicators to highlight the links between agriculture-environment, energy-environment, transport-environment and tourism-environment.

Eurostat has also been one of the key movers behind the development of environmental accounts, integrated with national economic accounts since 1995. Statistical handbooks and numerical results have been published in several areas of environmental accounting such as natural resources, air emissions, environmental expenditure and material flow accounts.

Eurostat also operates a data sharing policy with the EEA. Since 1999, the two organisations have had a Memorandum of Understanding that aims to:

> ... build on the existing strengths of both organisations and to avoid duplication of work, and that joint activities should be structured so as to build an efficient and transparent environmental information system.

For example, to avoid duplication of data requests to Member States, Eurostat uses data on emissions of air pollutants and greenhouse gases and on air quality collected by the EEA.

Decision flows

There are six directors of Eurostat, each responsible for different sectors of the organisation's activities:

(1) resources;

(2) statistical methodologies and tools;

(3) economic and monetary statistics;

(4) single market, employment and social statistics;

(5) agricultural, fisheries, structural funds and environment statistics; and

(6) external relations statistics.

In June 1997 Article 285 was inserted into the Amsterdam Treaty, providing Community statistics with a constitutional basis for the first time:

> ... the Council ... shall adopt measures for the production of statistics where necessary for the performance of the activities of the Community.

> The production of Community Statistics shall conform to impartiality, reliability, objectivity, scientific independence, cost-effectiveness and statistical confidentiality; it shall not entail excessive burdens on economic operators.

More and more statistics have to be collected at Community level because of the development of the EU, and it is no longer possible to ensure the availability of EU statistics on the basis of agreements alone. The Council of the EU adopted what is known as the Statistical Law in February 1997. This Regulation defines the division of responsibility between national and Community statistical authorities.

It also defines the basic conditions, procedures and general provisions governing official statistics at EU level. A Commission Decision later in 1997 clarified the role of the Community Statistical Authority – Eurostat – defined in the Council Regulation. Secondly, it reaffirmed the need for those involved in Community statistics to follow fundamental principles in ensuring that statistics are scientifically independent, transparent, impartial, reliable, pertinent and cost-effective.

Information flows

The ESS was built up with the objective of providing comparable statistics at EU level. The ESS comprises:

- Eurostat and the statistical offices, ministries, agencies and central banks that collect official statistics in EU Member States;

- international organisations including the OECD, United Nations Statistics Division (UNSD), the International Monetary Fund (IMF) and the World Bank; and

- the EEA.

The ESS functions as a European network in which Eurostat's role is to lead the way in the harmonisation of statistics in close co-operation with the national statistical authorities. The Statistical Programme Committee (chaired by Eurostat) brings together heads of Member States' national statistical offices in order to discuss the most important joint actions and programmes to be carried out to meet EU information requirements.

The current multi-annual statistical programme covers the period 2003–07. In this programme, the statistical implications of the major Community policies are examined in terms of major European initiatives such as Economic and Monetary Union, EU enlargement, competitivity, sustainable development, the social agenda and structural indicators.

Apart from the partnership with the national statistical systems of the Member States, co-operation will also extend to the systems in EFTA countries that form part of the European Economic Area, as well as to international organisations active in the statistical field, particularly the United Nations and its agencies and the OECD.

Organisational collaboration and standardised environmental data collection and monitoring within Europe raises a number of issues that Eurostat and other European organisations (including the European Commission DG Environment and the EEA) are currently addressing. A current focus is the integration and future use of geographical information and GIS for environmental analyses. At Eurostat, the Geographic Information Service of the European Commission (GISCO) is the provider of georeferenced data; environmental data is one of the pillars of the GISCO database, which comprises a series of multipurpose geographical data.

Funding

Eurostat has an operating budget of €141 million (2000) of which 50% supports the implementation of the Statistical Programme, 38% is staff costs, and the remainder is spent on administration. Additional income is generated through the sales of its products and databases. The Environment and Sustainable Statistics Unit has approximately 20 staff and an operating budget of €2 million, not including administrative overheads. GISCO has seven full time staff and has an annual operational budget of approximately €500,000. An extra €200,000 was received in 2003 for activities related to the preparation of the INSPIRE framework directive.

Eurostat and GMES

GMES has requirements for good socio-economic data for a number of thematic areas, namely air pollution, noise pollution, land use change, climate change and waste generation. Eurostat represents a key provider of this information, as well as extensive statistics on the state of the environment harmonised across European countries. Further harmonisation of reporting activities needs to take place between the European Commission DG Environment, the EEA and Eurostat, a process that has already been started and which can be boosted by GMES.

EUROPEAN UNION SATELLITE CENTRE

EUSC objectives

The EU Satellite Centre (EUSC) is dedicated to the exploitation and production of information from satellite images in support of the EU's Common Foreign and Security Policy (CFSP), and in particular for the European Security and Defence Policy (ESDP). In line with this mission, the EUSC provides information to the EU, Member States, the Commission, third states and international organisations for, amongst others, the following areas:

- general security surveillance;
- Petersberg missions, that is, support to humanitarian and rescue tasks, support to peacekeeping tasks, tasks of combat forces in crisis management, including peacemaking (Mattocks 2003);
- treaty verification;
- arms and proliferation control;
- maritime surveillance; and
- environmental monitoring, including both natural and man-induced disasters.

The EUSC's activities are focused on security, but besides the military aspects it also covers areas related to civil security. On the civil security side, it has a close working relationship with the European Commission's JRC and is developing activities for providing satellite imaging services for humanitarian agencies and partners of the UN on a not-for-profit basis. The relationship that the EUSC has with various partners is illustrated in Figure 10.5.

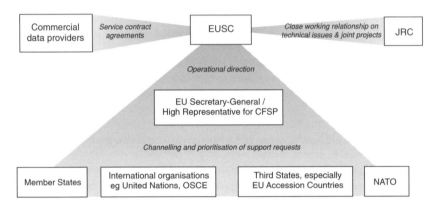

Figure 10.5: Key institutional links for the European Union Satellite Centre

Decision flows

The EUSC was created by a Council Joint Action (European Council 2001) on 20 July 2001, drawing on the assets and experience of the Western European Union Satellite Centre, which had been formed in 1993. The EUSC became operational on 1 January 2002 under the political supervision of the Political and Security Committee and the operational direction of the Secretary General. The Centre has its own legal framework in order to carry out its functions. It conducts imagery analysis training and undertakes research and development projects to enhance its capabilities and efficiency.

Information flows

The EUSC has established service contract arrangements with commercial Earth observation and aerial data providers to support their operational needs for access to digital imagery. The EUSC also works closely with the JRC to exchange technical expertise and resources in image processing and interpretation.

Funding

The EUSC is funded through contributions from the EU Member States and has an annual operating budget of €9–10 million. The Centre has 68 staff, including 15 image analysts. In some cases, the entire image analyst resource has been deployed to cover a single event. This is seen as a major constraining factor in support of the increasing ESDP commitments and the wide range of operational requirements of the security component of GMES.

The EUSC and GMES

The EUSC represents a key link for GMES in the area of security. It has experience in providing not only military information, but also information in support of civil security activities related to the Petersberg tasks as well as natural and man-made disasters. The EUSC could play a role in developing dual use mechanisms and procedures. The role of the EUSC is still developing with regard to Europe's new CFSP. The CFSP is, itself, also at a relatively early stage of development. Currently, the CFSP is considered the 'second pillar' of the EU, but lies outside of the authority of the EC and its scope is debated by Member States. A significant role for the EUSC in GMES may evolve in due course as the CFSP issues are resolved and a clear definition of security in the GMES context is finalised.

JOINT RESEARCH CENTRE

JRC objectives

The remit of the European Commission's JRC is to provide enhanced scientific and technical support to a range of EU policies. The JRC is directly funded by the Commission under a four year rolling framework budget for research and technology development. Its current multi-annual work programme strongly follows the Commission's strategic objectives with particular emphasis on enlargement, sustainable development, safe and clean energy, protection of the environment, food safety, health and security. Its activities or projects have end-users, in particular policy DGs, especially those in DG Environment and DG Agriculture, but also in Health and Consumer Protection, Regional Policy and the Humanitarian Aid Office. The JRC also has numerous links with other European and international organisations, as illustrated in Figure 10.6.

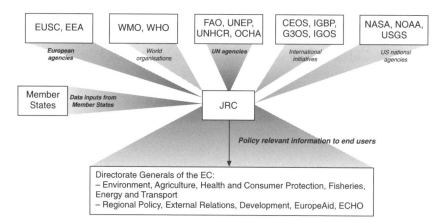

Figure 10.6: Key institutional links for the JRC, focusing on GMES-related activities

Projects carried out at the JRC make wide use of satellite Earth observation data, while *in situ* data is supplied by the individual European Member States. In both cases, this limits distribution of the resulting datasets that projects produce. Composite images are subject to satellite data provider licensing agreements and Member States often limit distribution of *in situ* datasets to internal EC use only. Where their datasets can be made freely available, this is done via the Internet, for example the Global Land Cover 2000 dataset (GVM 2004).

The JRC and GMES

The JRC has been a catalyst in the GMES process since it began and subsequently supports the development of GMES in a number of areas. Through the work with policy DGs, the JRC assesses and develops institutional demand for data and information services, identifies shortcomings in present monitoring infrastructures and checks the feasibility and effectiveness of proposed new solutions. Present JRC activities that are relevant to GMES include the following:

- Monitoring the global environment, through land cover assessments, sustainable forest management, biodiversity, ocean productivity and the atmosphere.

- Environmental policies with a European geographic focus, through water and air quality, land use change and forestry, urbanisation, soil condition, nature protection sites and the implementation of the EU's Kyoto reporting obligations.

- European civil protection, through monitoring of floods, fires, landslides and marine oil spill monitoring.

- The Common Agricultural and Fisheries Policies, through monitoring area-control measures, forecasting crop production and detecting fishing vessels.

- European Union external aid and security policies, through provision of mapping and decision support services for aid, reconstruction and de-mining, and development of tools for verification of non-proliferation treaties.

EUROPEAN CENTRE FOR MEDIUM-RANGE WEATHER FORECASTING

ECMWF objectives

The ECMWF is an international organisation supported by 24 European States. Its Member States are Austria, Belgium, Denmark, Finland, France, Germany, Greece, Ireland, Italy, Luxembourg, the Netherlands, Norway, Portugal, Spain, Sweden, Switzerland, Turkey and the UK. It has also concluded co-operation agreements with Croatia, Czech Republic, Iceland, Hungary, Slovenia, and Serbia and Montenegro, as well as the WMO, EUMETSAT, and the African Centre of Meteorological Applications for Development (ACMAD).

The principal objectives of the ECMWF are:

- the development of numerical methods for medium-range weather forecasting;

- the preparation, on a regular basis, of medium-range weather forecasts for distribution to the NMSs of the Member States;

- scientific and technical research directed at the improvement of these forecasts; and

- collection and storage of appropriate meteorological data.

The ECMWF generates operational forecasts (weather predictions up to 10 days ahead) for Member States and co-operating states, as well as to the WMO and the public. Outputs are also available to the research community through bilateral agreements such as those with the JRC and the British Atmospheric Data Centre (BADC). In addition, the ECMWF carries out research to support its forecasting activities. The Centre has an extensive network of links to other organisations on European and international levels. This is illustrated in Figure 10.7.

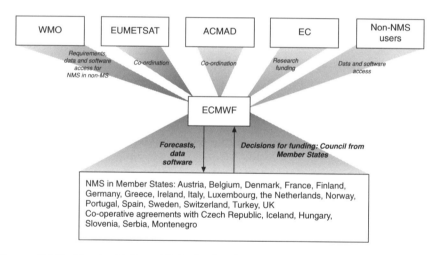

Figure 10.7: Key institutional links for the European Centre for Medium-Range Weather Forecasting

Decision flows

The ECMWF Convention defines the structure of the organisation:

> The organs of the Centre [the ECMWF] shall be the Council and the Director. The Council shall be assisted by a Scientific Advisory Committee and a Finance Committee.

Further Committees have since been established by the Council. The Council and Committees meet at regular periods throughout the year.

Information flows

All ECMWF model results are stored in the Centre's Meteorological Archival and Retrieval System (MARS). To be able to respond to requests for archived data, particularly for research and education purposes, ECMWF operates a Data Ordering Service. This service has been developed in accordance with the rules agreed by Council. Outputs are also available to the research community through bilateral agreements and through academic data centres such as the BADC.

ECMWF also provides its software and access to its computing facilities in accordance with the rules agreed by its Council. All national organisations designated by Member States and co-operating states are given a free non-exclusive licence to the software.

Funding

The annual budget of ECMWF is approximately €40 million. It has approximately 200 staff, of which about 75 are engaged in research activities. ECMWF is predominantly funded by its Member States according to GNI rules, updated every three years. Some extra funding is provided from sales of data and products.

ECMWF and GMES

The ECMWF is an operational entity with long-term public funding from Member States. It represents an important source of medium-term weather forecasts that constitute inputs into a number of environmental and security information topics, for example agriculture monitoring and climate change. It is at the forefront of research and development of assimilation schemes for weather forecasting purposes. ECMWF can be particularly helpful to GMES in the development of operational systems and in the technical approaches to environmental prediction.

EUROGEOGRAPHICS

EuroGeographics is the association of the European National Mapping and Cadastral Agencies (NMCAs) with 43 members from 40 countries – from Iceland to Turkey and from Portugal to Russia. The vision of EuroGeographics is to achieve interoperability of European mapping and other geographical information to underpin Europe's Spatial Data Infrastructure (ESDI).

EuroGeographics is funded by contributions from its members. Its administrative budget is around €500,000 per year for five permanent staff located at the head office near Paris, while additional funding is made available for specific project work.

EuroGeographics is a not-for-profit organisation. The work programme is overseen by a President, and a Management Board made up of representatives from the NMCAs. It operates as a networked organisation with permanent expert groups and with individual NMCAs taking responsibility for a project and conducting operations from its offices.

EuroGeographics has produced the following pan-European geographic information ('reference') datasets:

- SABE (Seamless Administrative Boundaries of Europe) – an administrative boundaries dataset at map scales of 1:100,000 and 1:1 million.

- EuroGlobalMap – a 1:1 million topographic dataset that will be the European contribution to the Global Map project.

- EuroRegionalMap – a 1:250,000 scale topographic dataset.

EuroGeographics is currently developing processes to open up access to other (larger scale) reference datasets, including road and river networks, cadastral information, height (digital elevation models) and geographical names. These technical developments are supported by work on business models, and on pricing and licensing policies to improve access to information, and by the development of partnerships with other pan-European associations, academia and the private sector, including value added resellers who distribute EuroGeographics' products to the marketplace.

The EuroGeographics' Geographical Data Description Directory has information on the availability of spatial data in Europe from the various NMCAs. For each country, a list of available map datasets is described regarding dataset specification, technical overview and commercial details. Contact details are also supplied for acquisition and purchasing.

EuroGeographics has an interesting and potentially influential role to play in GMES by ensuring that the necessary underpinning geographic framework is in place to which all other environmental and security information can be referenced, enabling integration and sharing of data across disciplines.

EUROGEOSURVEYS

EuroGeoSurveys, the Association of Geological Surveys of the European Union, was legally registered in France in January 1996 as a non-profit association. Its 25 members are the national geological surveys of all present EU Member States plus some of the candidate Member States. EuroGeoSurveys has defined its mission as follows:

- To jointly address European issues of common interest.

- To promote the contribution of geosciences to EU affairs and action programmes.

- To assist the EU to obtain technical advice from the members of the Association.

- To provide a permanent network between the Geological Surveys and a common, but not unique, gateway to each of the Surveys and their national networks.

EuroGeoSurveys' business is managed by an Executive Committee of four of the national directors, elected by the general meeting of all individual Geological Survey Directors. The general meeting governs EuroGeoSurveys and directs its forward policy. The Association mandates a full-time Secretary General as its legal representative at European level through a bureau based in Brussels, and who is supported by two staff members.

For its technical support to the European Commission, EuroGeoSurveys runs a number of expert groups in parallel with the legislative agenda of the Commission, for example the Groundwater Expert Group, the Soil Protection Expert Group and the Sustainable Development Expert Group.

EuroGeoSurveys is based on its staff in Brussels co-ordinating a network of over 8,000 geosciences experts across Europe, that is, the staff of the national geological survey organisations. As the technical input is limited to on-call and voluntary contributions, the total annual budget of EuroGeoSurveys is limited to around €400,000, 90% of which consists of membership contributions, and 10% is direct cost recovery for expert advice to the European Commission.

The national geological surveys maintain large geoscience information banks, built from information gathered continuously over periods of up to 170 years. As these datasets come from different sources and are from different vintages, their formats, measuring techniques and collecting philosophies differ largely, not only in time and space, but also in detail and quality.

In order to provide Europe with standardised, fully integrated data, EuroGeoSurveys and its members developed two major data centres with financial support from the European Commission. These are the Geological Electronic Information Exchange System (GEIXS) and the European Union Sea Sediments Database (EU-SEASED). EU-SEASED comprises of three Internet databases that have been merged into a single integrated portal. The three databases are:

(1) EUROCORE: a searchable database of seabed samples from the ocean basins held at European institutions, universities and marine stations;

(2) EUMARSIN: European Marine Sediment Network, concerning marine sediment databases of the geological survey of Europe; and

(3) EUROSEISMIC: European Marine Seismic Metadata Information Centre, combining all the publicly available seismic data gathered across European sea basins.

There already exists a good match between GMES and one of the objectives of EuroGeoSurveys, namely to provide input to EU policy formulation regarding environment, energy, development co-operation and regional policies. In addition, the two EuroGeoSurveys data centres hold information concerning geology and seabed and marine sediment, which will be of value to GMES.

EUROPEAN-MEDITERRANEAN SEISMOLOGICAL CENTRE

The EMSC was founded in 1975 and originally started its operations at the Institut de Physique du Globe de Strasbourg. It formally received its statutes in 1983 and in 1993 was moved to the Laboratoire de Détection et de Géophysique in Essonne, France. In 1987, the EMSC was charged by the Council of Europe to provide the Council with seismic warnings in the framework of the open partial agreement on the prevention of, protection against, and organisation of relief in major natural and technological disasters.

The EMSC is an international, non-governmental, and non-profit association. It plays a federating role at the European level. It has an annual budget of approximately €125,000, its main contributor being its members. EMSC members are mainly seismological institutes. Members include EU Member State organisations but also organisations from countries bordering the Mediterranean including Albania, Algeria, Croatia, Cyprus, Egypt, Israel, Lebanon, Monaco, Morocco, Serbia and Montenegro, Slovenia and Turkey, as well as other countries such as Armenia, Belarus, Bulgaria, Czech Republic, Georgia, Iceland, Romania, Russia, Saudi Arabia and Slovakia. In 2003, there were 57 members in 37 countries, but the networking activities involve more than 100 institutes.

The EMSC runs an Earthquake Warning System for potentially damaging earthquakes, which focuses on the European-Mediterranean region. The system consists of the rapid determination of the epicentre and magnitude of an earthquake and the dissemination of the seismic alert message within the hour following the occurrence of the earthquake. This operational alert system – known as the European Alert System – is triggered by any earthquake whose magnitude is greater than 5.0 (Richter scale) over the European-Mediterranean region, greater than 5.5 in the Mediterranean basin and greater than 7.0 worldwide. The EMSC maintains a continuous operational activity under the control of the seismologist on duty with the support of its host institute, Laboratoire de Détection et de Géophysique in France, and back-up procedures at Instituto Geografico Nacional (IGN, Spain). More than 40 seismological station networks contribute data, in quasi-real time, related to seismic activity to this EMSC alert system.

It is important to note that the alert messages are generally not sent to national civil protection agencies, as they are informed by their national and authoritative institutes. The interest of this system is its geographical coverage, as it provides a back-up for national institutes, and ensures information dissemination outside the affected country.

The operational alert system run by EMSC in the European-Mediterranean region is relevant to GMES in a number of ways. On the one hand, it offers valuable input to the preparedness, rescue and crisis management tasks related to earthquakes both within Europe and for many countries outside the EU. It also illustrates a good model for an NRT operational service that is operated on a continuous basis using data that is collected via a network of stations located across member countries. In addition, one of the scientific objectives of the EMSC is to make this data (collected for operational use) available to other international, regional and national data centres, and to promote data exchange between laboratories in the European-Mediterranean region.

CONCLUSIONS

In this chapter we have reviewed the key European organisations in light of what they could bring to the GMES initiative. These have included:

- European Commission Directorates, the JRC and Eurostat;
- agencies of the EU such as the EEA and the EUSC;
- intergovernmental organisations such as ESA, EUMETSAT and ECMWF; and
- non-governmental organisations such as EuroGeographics, EuroGeosurveys and EMSC.

Between them, they hold a wide range of key competencies corresponding to specific tasks in the system of production of information on the environment and security at the European level, from satellite build to state of the environment reporting, from research and development to fully operational service provision. Under a GMES scenario, the key European actors will need to work together with the aim of achieving the capability to effectively monitor the environment and ensure civil security.

At Member State level, systems for the production of environmental and security information have been implemented in response to national needs, including compliance with international treaties and European directives and policies. A complex web of systems and services already exists, albeit with substantial duplication and gaps at a European level. The need for improved co-ordination is well understood and many of the organisations described in this chapter (for example, EuroGeoSurveys, EuroGeographics, EMSC) are already working towards achieving improved co-ordination within Europe and its neighbouring states. To facilitate further improvements in collaboration and sharing of data and information, many of these organisations will need to review their data policy, data access and sharing arrangements outside their own communities.

Each of these organisations was established with a specific remit in mind, and in some instances this remit has been broadened over time (for example, the EEA and EUMETSAT). It may be necessary that this is widened further, or extended to other organisations, for the jigsaw puzzle of GMES to be completed. In all cases this will require agreement from the individual governing boards or councils of these organisations.

Issues concerning the environment, as well as security issues, are not constrained by geopolitical boundaries. Therefore, by its nature, GMES will involve countries outside the EU. This is already reflected in the membership of many of the above organisations, including neighbouring accession countries, Balkan states and countries situated around the Mediterranean.

Weather monitoring and forecasting in general represents a good model for GMES to follow regarding the structures and organisations that are needed to deliver operational environmental services. The seven key characteristics of the meteorological system are listed below:

(1) It operates at a:

- global level (through the WMO and its committees, for example the Co-ordination Group for Meteorological Satellites);

- European level (for example, EUMETSAT, ECMWF, and the European Meteorological Services Network, EUMETNET);

- national level (through NMSs); and

- local level (for *in situ* data collection).

(2) Day-to-day business processes, driven by the need to produce a new forecast every six hours, and key decision-making processes (such as requirements capture and translation into observational needs, funding and legal frameworks) are all in place.

(3) The global observational system defined by the WMO includes both *in situ* and satellite data. The meteorological community provides an important example of how a range of data sources can be integrated successfully.

(4) Driven by the need for improved forecasts, sophisticated models and numerous data sources are combined within state of the art computing systems. There is a strong relationship between the operational community and the research community, and both are funded.

(5) While each Member State produces its own forecast, it relies heavily on sharing data from other NMSs and satellite operators. The resulting architecture is highly distributed with data collection at the appropriate level. NMSs make their data available to others over the Global Telecommunications System (GTS). The need to share data routinely in real time has also led to carefully defined standards, data exchange formats, etc, and co-ordination activities such as EUMETNET.

(6) While it is currently appropriate for each country to produce its own forecasts, for national policy requirements, certain activities are undertaken at European level to avoid unnecessary duplication and reduce costs (for example, procurement and operation of satellites, production of medium-range forecasts).

(7) The system is able to accommodate a wide range of users – research, public, government and commercial.

The other subject areas of GMES, such as air quality, land use, land degradation, marine pollution or civil security, are less mature with respect to delivering operational services. The challenge for GMES lies in developing such operational services to deliver information on the wide range of topics within the spheres of environment and security.

GMES is a complex undertaking and thus a more formalised structure might be considered to ensure collaboration between the relevant organisations and the effective use of available resources to provide operational services. One such structure that has already been used at a European level is a Joint Undertaking. Such an undertaking has recently been formed between the European Commission, ESA and, potentially, outside investors (for example, banks or private industry) for the deployment and operational phases of the Galileo Programme (Europe's global satellite navigation system). A Joint Undertaking has the added advantage that it can be funded by means of a budget aggregated from various budget lines from within the organisations involved.

CHAPTER 11

SOCIO-ECONOMIC BENEFITS: ANALYSIS OF GLOBAL ENVIRONMENTAL MONITORING

Pam Vass

INTRODUCTION AND METHODOLOGY

Structure

Ultimately, any form of global environmental monitoring will stand or fall on the benefits it brings to the organisations that provide the funds. This chapter explores the potential socio-economic benefits of global environmental monitoring in general, and the Global Monitoring for Environment and Security (GMES) programme in particular, by reporting on a research project carried out as part of the initial phase of GMES during the year 2003. This chapter analyses socio-economic benefits through the following six studies:

(1) Natural disasters and risk reduction.

(2) Monitoring of earthquakes and volcanic eruptions (geohazards).

(3) Oil spill detection.

(4) Ocean monitoring to improve weather forecasting.

(5) Air quality monitoring.

(6) Climate change research.

Evidence from the literature to support the information is cited in the text: extra information that is not cited directly is included in Appendix B. The main findings of the research are summarised along with suggestions for prioritisation. Before launching into the six case studies in detail, the next sections introduce the foundations behind social, economic and environmental evaluations of benefits.

Drivers and foundations

The justification for investing public funds in global environmental monitoring programmes, systems or frameworks is generally made on the basis of four different drivers. First, *Saving the Planet*. The case assumes that the risk of damage to the planet from global warming, extreme weather, the loss of biodiversity, etc, is so high that environmental information and monitoring services to protect the future of the human population and to detect and warn of adverse change are essential. Given the high stakes and the need to save the planet, there is little point in arguing about the cost. Secondly, *Sustainable Development*. Conventions and other international agreements make it obligatory for nations to manage and control the exploitation of the environment in such a way that conservation and productivity are preserved. Thus, a methodology has to be used to measure and compare costs, benefits, losses and damage. Thirdly, *True Environmental Value*. The natural environment represents Environmental Capital or Ecosystem Services (Costanza *et al* 1997, Harris 2002), and performs valuable functions such as keeping drinking water clean, or providing the right balance of oxygen and carbon dioxide in the atmosphere, which we would have to pay for if nature did not do it for us. Although not without estimation problems, the value of these services can and should be included or acknowledged in financial calculations of the benefits of global environmental monitoring. Fourthly, *Commercial Economic and Social Value*. A full Cost Benefit Analysis (CBA), as performed for a commercial enterprise, can be used to evaluate the costs and benefits of having better information.

Scope of socio-economic benefits analysis

Most socio-economic benefit studies address only the fourth driver to estimate the (commercial) benefits of an activity. To do this well, such CBAs depend upon identifying a mature service, its reliance upon environmental information or forecasts, the likely up-take of information, the correct and reliable use of forecasts, and the difference between the efficiency or profit of the activity after using the information, compared with the performance without the information. This has proven to be a difficult task, if not impossible, even for single industries or services in limited geographical regions. Thus, the usual approach is to hypothesise that the programmes or frameworks could improve turnover or profit of the activity by 1–2% of value added.

A more sophisticated analysis is clearly preferable. The route for programmes such as GMES in the future will be to perform an Extended Impact Assessment (ExIA) that will be able to take into consideration non-quantitative costs and benefits as associated with issues such as 'saving the planet', 'sustainable development' and 'environmental capital', as well as detailed economic, social and environmental benefits and costs. At this early stage in the development of GMES, before the full scope of GMES has been established, it was not appropriate to perform an ExIA and a more simple approach has been taken. The factors that provided the constraints or boundary conditions to the socio-economic benefits analysis are listed below:

- The assessment does not include any costs associated with the development of GMES itself.

- The task has been based on information from published reports and budgetary figures by sector or environmental theme.

- The assessment concentrates on quantitative benefits of improved environmental monitoring, and not on non-monetary qualitative benefits.

- The assessment concentrates on benefits to Europe and only in some cases considers the wider role Europe plays throughout the rest of the world.

Although it is possible to identify cases of inefficient use of budgets resulting in monetary and resource wastage, that is, cases where monitoring networks do not collect the optimum data either in type or coverage, these inefficiencies vary from country to country and topic to topic and thus each would require a detailed survey. However, most of the resultant costs and savings of redressing these issues would be borne by and accrue to Member States at the national level. Further, when such savings are escalated to the European-wide scale they are of the order of hundreds of thousands or perhaps a few million euros and it would take many such cases to provide a financial case for GMES. So, although these issues have not been ignored, more attention has been directed towards ascertaining potential roles of GMES that will result in socio-economic benefits of hundreds of millions of euros. Such social, economic and environmental gains are likely to result from activities beyond the scope of individual Member States and on themes that we are only just beginning to address, but where collaborative ventures will further our understanding of the European and global environment and address civil security issues of high social priority.

The basis of economic evaluations

Each of the case studies reported in this chapter required socio-economic evaluations to be performed in different ways. The following eight methods were used to assist with the socio-economic benefits evaluations:

(1) Ascertaining market sectors worldwide and in Europe and applying a very low percentage to indicate the possible saving that improved monitoring would realise – a top-down macro-estimate.

(2) Ascertaining monitoring budgets for individual services in individual countries and multiplying up for the Europe-wide situation, then applying a low percentage to provide an indication of the possible saving of improved environmental monitoring. Often these figures have been checked with relevant budgets in the US for comparative purposes. This type of valuation is identified as a bottom-up micro-estimate.

(3) The translation of US costs and benefits to the European stage, taking account of the differences in Gross Domestic Product, area and population.

(4) Determining expenditure on EU projects for particular environmental themes and proposing savings through improved access to data, information, tools, etc. This implies that important project results are not lost to the research community and duplication in project efforts in the various Directorates General (DGs) and framework programmes is avoided.

(5) Initial cost benefit studies performed in the US as part of the Integrated Sustained Ocean Observing System (ISOOS) preparations and other studies such as Global Ocean Observing System (GOOS) and EuroGOOS, the economic valuation of improved weather forecasts in the US and Canada and the economic impacts of drought.

(6) The calculation of social costs, based on willingness to pay estimates, was based on work undertaken in Europe on air pollution using figures of €900,000

per life saved and €300,000 per life unaffected. The latter figure (which is health-related) may not be appropriate for all disaster types or worldwide estimates, but it provides a starting point for further, more detailed assessments.

(7) Consulting the work performed by the Framework Definition and Support (FDS) Working Group of the Infrastructure for Spatial Information in Europe (INSPIRE) (EC 2003b).

(8) Addressing a few one-off projects to determine budgets in terms of costs or effort as an indication of savings from sharing data rather than repeating tasks or having easier access to information.

It may be considered that this kind of analysis contains a certain degree of 'noise' and uncertainty in assumptions that will not stand up to a sceptical view. However, the methods provide orders of magnitude for comparative purposes and a baseline for more rigorous assessment in future global environmental monitoring programmes.

Using scenarios

In a number of the case studies given below, a range of figures are presented in the form of scenarios. The scenarios are models to allow estimates of future benefits. Scenario 1 is a simple improvement in monitoring capability through better access to data and information, and with an emphasis on characterising environmental events and issues. Scenario 2, with an emphasis on understanding events, permits a more comprehensive integrated service, bringing together diverse but complementary data sources and stakeholders. Scenario 3 is a fully integrated monitoring service with forecasting and prediction capabilities.

Scenarios 1 and 2 are seen as simple progressions in time in the evolution of a service, although different themes will start at different positions along that time frame. Scenario 3 is a marked evolutionary stage as it represents the transition to services based on forecasting and prediction (from reactive to proactive monitoring). These can be mapped against a GMES timeline and an effectiveness versus effort scale. Figure 11.1 summarises the scenarios and the possible time line.

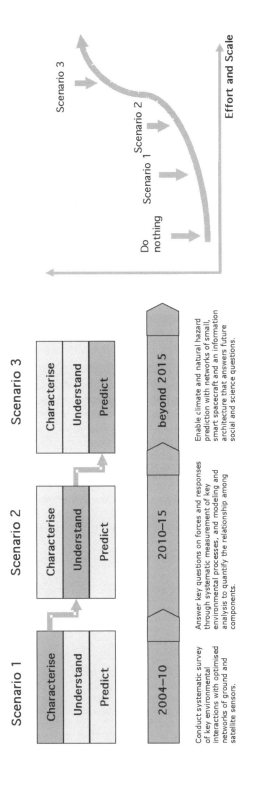

Figure 11.1: Economic valuation scenarios

NATURAL DISASTERS AND RISK REDUCTION

Cost of natural hazards

The period from 1970 to 1999 was one of an increase in the number and effects of disasters. Table 11.1 shows for the last three decades of the 20th century the number of reported disasters, the number of people killed and otherwise affected by the disasters and the economic losses sustained.

	1970–79	1980–89	1990–99
Number of reported disasters	1,110	1,987	2,742
People killed by disasters	1,960,000	800,000	790,000
People reported affected by disasters	740,000	1,450,000	1,960,000
Economic losses (year 2000 values)	US$131 billion	US$1,987 billion	US$2,742 billion

Table 11.1: Thirty years of natural disasters. Source: MunichRe Topics 2002

Table 11.2 indicates average annual disaster figures and estimates of social impacts and costs of various natural disasters for the period 1992–2001. These figures are probably an underestimate. Famine reportedly killed 280,000 people over this 10 year period, but well-placed sources estimate that those who died in the famine just in the Democratic People's Republic of Korea from 1995–98 may have numbered between 800,000 and 1.5 million (IFRC&RCS 2002). Of all those killed by natural disasters, 83% were Asians. On average, natural disasters accounted for 88% of all deaths from disasters over the last decade.

In general terms, between 1970 and the end of the 20th century, deaths from natural disasters fell from two million to under 800,000, but the numbers affected tripled to two billion and economic losses multiplied five times, to US$629 billion (€566 billion) in the 1990s.

Phenomenon	Number and economic valuation of people killed		Number and economic valuation of people affected		Estimated damages (€ million)		TOTAL (€ million)	
	Europe	World	Europe	World	Europe	World	Europe	World
Avalanches and landslides	103 €93	946 €851	1,712 €514	212,784 €63,835	€2	€131	€609	€64,818
Droughts and famines	0 €0	27,757 €24,982	601,000 €180,300	44,593,574 €13,378,072	€859	€2,664	€181,159	€13,405,717
Earthquakes	2,100 €1,890	7,776 €6,998	243,634 €73,090	3,494,214 €1,048,264	€2,283	€18,733	€77,263	€1,073,995
Extreme temperatures	254 €228	1,013 €912	75,399 €22,620	627,756 €188,327	€84	€1,104	€22,933	€190,342
Floods	136 €123	9,651 €8,686	572,043 €171,613	126,187,973 €37,856,392	€2,759	€15,177	€174,494	€37,880,255
Forest and scrub fires	15 €14	57 €52	12,332 €3,700	342,105 €102,632	€17	€2,035	€3,730	€104,719
Volcanic eruptions	0 €0	26 €23	0 €0	104,252 €31,275	€2	€39	€2	€31,337
Windstorms	71 €64	6,045 €5,440	729,698 €218,909	24,519,790 €7,355,937	€1,395	€14,157	€220,368	€7,375,534
Grand Total	2,679 €2,411	53,271 €47,944	2,235,818 €670,745	200,082,447 €60,024,734	€7,401	€54,040	€680,558	€60,126,718

Table 11.2: Average annual disaster figures and estimates of social impacts and costs by phenomenon for the period 1992–2001. Source: World Disasters Report (IFRC&RCS 2002). The top figure in columns 2 to 5 is the number of people and the bottom figure is the economic valuation (in € million) based on standardised 'willingness to pay' estimates of €900,000 per life saved and €300,000 per life unaffected (UNECE 1996)

Economic valuation of the benefits of GMES

Table 11.3, based on the three scenarios outlined earlier in this chapter, indicates what benefits GMES may bring to the monitoring of this range of natural hazards.

The data in Tables 11.2 and 11.3 indicate that during this period the major disasters in Europe and the rest of the world, in priority order and based on an economic evaluation, were as follows:

- Europe: floods, earthquakes, windstorms, droughts.

- Rest of the world: floods, droughts, famine, windstorms.

These figures and prioritisations relate to humanitarian costs and related damages and do not include full economic or insurance losses. For example, the forest fires in Portugal in 2003 claimed 18 lives and it is estimated that 270,000 hectares of forest and 25,000 hectares of agricultural land were destroyed and the disaster caused the loss of goods, jobs and employment for about 45,000 persons. Figure 11.2 shows the average annual area destroyed by fire in Europe, indicating that the 2003 summer fires in Portugal were markedly high compared to the sampled time period of 1988 to 1992. Using an average of the lumber value for Portugal (€9,000–€40,000 per hectare, summer 2000) the economic loss of forestry (assuming it was at a harvestable stage) would be in the region of €7 billion, again considerably more than the total economic valuation of fires for the whole of Europe of €3.7 billion (see Table 11.3).

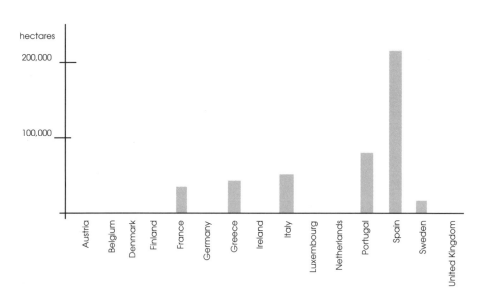

Figure 11.2: Area destroyed annually by fire in Europe 1988–92, average in hectares. Source: OECD

€million	European Risk Reduction			Europe's Contribution to World Risk Reduction		
	Scenario 1	Scenario 2	Scenario 3	Scenario 1	Scenario 2	Scenario 3
Avalanches/landslides	0.02	0.11	325	0.44	2.19	10,786
Droughts/famines	8.59	42.96	176,069	8.88	44.39	2,233,931
Earthquakes	22.83	114.15	265,780	62.44	312.21	176,501
Extreme temperatures	0.84	4.22	19,862	3.68	18.39	31,577
Floods	27.59	137.94	361,748	50.59	252.95	6,311,352
Forest/scrub fires	0.17	0.85	3,560	6.78	33.92	17,182
Volcanic eruptions	0.02	0.09	176	0.13	0.64	5,218
Windstorms	13.95	69.74	248,968	47.19	235.95	1,227,368
Grand Total	74	370	1,076,490	180	901	10,013,914

Table 11.3: Economic valuation of the potential benefits of various levels of monitoring (in € million). The following multipliers have been used: 1% of the annual estimated damages for Scenario 1; 5% of the annual estimated damages for Scenario 2; and 10% of the estimated damages and 50% of the economic valuation associated with reducing the number of people killed and people affected for Scenario 3.

Key findings

From the analysis of the costs of natural hazards and the related work on social and economic benefits of risk reduction, it is possible to identify the key findings and the factors that influence risk reduction in relation to natural hazards:

- Disasters are complex problems, which demand complex responses. Since risk reduction goes to the heart of the development process, the challenge is well beyond the capacity of disaster managers alone. It requires co-operation between development agencies, governments, NGOs, businesses, scientists and vulnerable communities.

- One factor increasing vulnerability is the absence of effective disaster preparedness and mitigation measures (for example, early warning systems, identification of evacuation routes and location for shelters, disaster response teams, public awareness).

- There are many documented success stories which show that mitigation and preparedness do pay.

- There is no coherent risk reduction community, as professionals trying to mitigate disaster impacts are fragmented along institutional and national boundaries.

- Data on disaster occurrence, its effect upon people and its cost to countries remains, at best, patchy. No single institution has taken on the role of prime providers of verified data, although the Centre for Research on the Epidemiology of Disasters (CRED) is a leader in this sector.

GEOHAZARDS MONITORING

Cost of geohazards

Geohazards such as earthquakes, volcanic eruptions, landslides and subsidence inflict an enormous cost on society (see Table 11.2). Every decade thousands of people are killed, yet this is only part of the toll: for every life lost, many more are injured, or lose their homes or livelihoods. A major disaster disrupts the economic life of a society for years or decades. Even where loss of life is avoided, geohazards damage infrastructure, destroying roads, railways, buildings, airports, pipelines, dams, power grids and many other structures. The cost of these events is valued at billions of euros and they affect the richest and poorest countries alike. As a consequence, private organisations most exposed to these risks seek to insure against them at an additional cost that is itself in the billions of euros. The UN has established that total insurance costs have risen tenfold in the past 40 years. The increasing risks posed by geohazards to all societies require better understanding of the hazards and better means to deal with them.

In addition to the social and economic costs of the event itself, longer-term effects often follow in the wake of an event – for example, from disease, water shortages and food shortages. Furthermore, volcanoes may have significant impacts on climate and air pollution. Even the low magnitude eruptions of Hekla in Iceland in February 2000 caused US$3 million (€2.7 million) of damage to the engines of a NASA aircraft that encountered the eruption plume (Grindle and Burcham 2002).

Benefits of monitoring

Many of the last decade's significant earthquakes occurred on faults that had not previously been recognised as major faults in their regions (for example, Northridge, California, 1994; Kobe, Japan, 1995; Athens, Greece, 1999). It is here that improved monitoring and, in particular, satellite technology could have a big impact, by mapping strain accumulation globally and producing reliable medium-range earthquake forecasts. Wright (2002) argues that the economic case is simple. The 1994 Northridge earthquake in Los Angeles caused total property damage estimated at US$20 billion (€18 billion). This would have been greater were it not for an intense programme of hazard mitigation activities over the previous two decades. Rebuilding or retrofitting structures to protect them from earthquakes is relatively cheap compared with the cost of rebuilding after an earthquake. For example, the US Federal Emergency Management Agency (FEMA) estimates the cost of retrofitting bridges to be just 22% of the cost of rebuilding if they are destroyed by earthquakes, and this does not take into account the cost to the local economy of the temporary loss of infrastructure.

Protecting against earthquakes is an issue that involves engineers, planners, politicians and the insurance industry. However, to act effectively, the civil and business communities need information with which the hazard from earthquakes can be quantified, and this information needs to be provided by the scientific community. It is a matter of great importance, therefore, that there should be continuing dialogue and co-operation between these two communities. The business world needs to benefit from the latest scientific understanding of the processes controlling seismic activity, but this can only be done if they can receive that information in a form that is relevant to their needs.

Economic valuation of the benefits of GMES

With regard to what benefits GMES may bring to this area of environmental security monitoring, Table 11.3 indicates the economic valuations of improved monitoring of earthquakes and volcanic eruptions for the three improved monitoring scenarios for Europe, and for Europe's contribution to the rest of the world (one third of total world figures minus European figures). Summing the European and world figures together gives total annual benefits (rounded estimates) for earthquake monitoring of €85 million for Scenario 1, €426 million for Scenario 2 and €442 billion for Scenario 3. Scenario 3 would require optimised ground networks (see below) and operational Interferometric Synthetic Aperture Radar (InSAR) capabilities. For volcanic eruption monitoring, the total annual benefits are €150,000 for Scenario 1, €750,000 for Scenario 2 and €5.4 billion for Scenario 3. However, in a detailed study each hazard type would be treated differently. For example, improved monitoring of the more predictive nature of volcanoes could be expected to see higher benefits earlier.

As regards the current national seismic monitoring systems, it is estimated that the total budget for the 25 members of the EU (the EU25) may be in the order of €100 million (based on the British Geological Survey figures in 2002). This compares well with a total US budget (2002) of US$73 million (€66 million) for general seismic monitoring. Much of the European network is reported as being outdated and, having evolved over a relatively long period of time, may not now represent the best network for European countries working together. In addition, it is known that there are major gaps in the network. Thus, although there will be national expenditures to update the network, the resulting upgrade, if done correctly, will eliminate duplication of effort while providing the highest quality data that serves multiple purposes for integrated monitoring. Following completion of such a network, it could be anticipated that a saving of at least 10% (approximately €10 million) would result from annual European geohazard monitoring activity.

Key findings

From the analysis of the costs of geohazards and the related work on social and economic benefits, the list below identifies the key findings and the factors that influence geohazard risk:

- Earthquakes are by far the most destructive natural disasters and they are one of the main natural hazards in Europe.

- A third of the world's population lives in areas considered to be at risk from geohazards, thus the potential for significant losses increases with population growth, urban expansion and densification, and the spread of industrial and infrastructure development.

- Major volcanic eruptions do not occur spontaneously and are preceded by a variety of physical, geological and chemical changes. The monitoring and measurement of these changes with well-established scientific techniques provide the best opportunity to develop an early warning system.

- All European countries have operational seismic networks and services, but much of the European network is reported as being outdated and, having evolved over a relatively long period of time, may not now represent the best network for European countries working together. In addition, it is known that there are major gaps in the network. However, mobile networks can be used to fill gaps in the short term.

- There are relevant European scientific organisations, including the European Seismological Commission (ESC), the European-Mediterranean Seismological Centre (EMSC) and the Observatories and Research Facilities for EUropean Seismology (ORFEUS). There are relevant European humanitarian organisations, including the European Commission Humanitarian Aid Office (ECHO) and the Centre Européen de Prévention des Risques (CEPR), France.

- A number of the separate elements for seismic hazard monitoring exist, but the topic will benefit considerably from integrated efforts and an overarching organisation along the lines of FEMA.

- A dedicated InSAR mission in the next 10 years, and perhaps a constellation of Earth monitoring satellites in the next 25 years, would lead to a vastly improved understanding of the physics of the earthquake cycle, a complete time-varying map of the Earth's strain and reliable earthquake forecasts that will save lives.

OIL SPILL DETECTION

Economic cost of oil spills

It is difficult to cost the impact of marine oil spills accurately – large spills occur relatively infrequently, perhaps one every three to four years. The most studied spill has been that from the *Exxon Valdez* in 1989, the worst spill in US history. The spill ended up costing US$2.1 billion (€1.89 billion) in clean-up expenditure and compensation – the vast majority of it paid by the Exxon oil company after a series of court cases. The spill killed 250,000 sea birds, 2,800 sea otters, 300 harbour seals, 250 bald eagles, up to 22 killer whales, and billions of salmon and herring eggs. The studies of sport-fishing activity and tourism indicators (vacation planning, visitor spending and cancelled bookings) all indicated decreases in recreation/tourism activity. A contingent valuation study estimated the lost passive use value at US$2.8 billion (€2.52 billion) (Carson *et al* 1992).

Enforcing stricter requirements on activities has led to a global decline of oil pollution inputs in the marine environment. In 1981, oil transportation and shipping in general were responsible for discharging about 1.4 million tons of oil products (Patin 1999). This amount was reduced to 560,000 tons in 1990. The reduction mainly occurred as a result of adopting stricter international regulations concerning transportation operations in the sea (for example, the International Convention for Prevention of Pollution from Ships). The total oil pollution input into the sea during the same period dropped from 3.20 to 2.35 million tons. This progress should continue with the eventual decommissioning of all single-hulled oil tankers in 2015.

Economic valuation of the benefits

Based on the UK annual marine pollution control budget of €1.7 million and an allocation of 50% to oil incidences, it is estimated for the 13 maritime EU15 nations that the total European budget for oil spill monitoring may be around €10 million per annum. If an oil spill detection monitoring service within a GMES framework resulted in economic savings of around 2.5–5% per year, this would produce a benefit of €250,000–500,000 per year. With regard to controlling oil spill damage, up-to-the-minute response information on currents and winds is essential for effective deployment of oil spill containment and clean-up efforts. It has been calculated that a 1% increase in the efficiency of oil spill clean-up would have saved the states in New England US$7.5 million (€6.75 million) over the last 10 years and nearly US$100 million (€90 million) in the US over the same period (Adams *et al* 1995).

Key findings

From the analysis of the costs of oil spills and oil pollution and the related work on social and economic benefits, the list below identifies the key findings and the factors that influence oil pollution risk:

- More oil enters the sea from land-based sources than tanker transportation and accidents.
- Large economically important spills are relatively rare.
- Regulations are leading to a global decline in oil pollution inputs to the marine environment.
- Difficulties exist in applying the polluter pays principle and effective prosecution of illegal discharges requires changes in national and European laws.
- Insurance generally covers clean-up expenditure and compensation.

It is difficult to identify substantial economic or social benefits that would result from an enhanced investment in marine oil spill detection monitoring services, but the benefit of GMES could be in the range of at least €1 million per annum. Following a spill, there are clear logistical roles that environmental data can offer in modelling the spill, targeting spraying, etc. However, this would be covered by a disaster risk reduction activity rather than a detection monitoring service.

OCEAN MONITORING

Economic valuation of the benefits

In the technically developed countries, marine resources and services contribute, on average, 5% of gross national product (GNP) (GOOS 1998). Thus, for the EU15 this is approximately €384 billion per year. The economic enhancement of this valuation by an improved monitoring service (based on a valuation of 1%) may be in the order of €4 billion per year. This is broadly in line with estimates by Flemming (2001) and Woods *et al* (1996) for the sum total of benefits by EuroGOOS for the European region being in the order of €2–5 billion per year. The Woods *et al* figure was primarily based on commercial economic values and did not include any environmental values or issues of sustainability. Furthermore, the marine information systems market is estimated to be valued at around €1.86 billion worldwide, with projected growth of between 7% and 12% per year (BMT 2001). Using this figure, a growth rate of 10% and an allocation of one-third to Europe, it could be estimated that the market for marine information systems in 2008 would be €1.2 billion in Europe.

Looking at a fully integrated ocean observing system with forecasting and prediction capabilities (Scenario 3), the potential economic and social benefits are much higher, and work in the US on the economic assessment of the Integrated Sustained Ocean Observing System (ISOOS) can be used to indicate anticipated benefits. Adams *et al* (1995) review the full range of benefits expected from ISOOS. Some of the benefits in non-marine fields are discussed below, while in relation to the value of improved forecasts from ISOOS to maritime transportation, commercial fishing, offshore energy production, defence organisations and search and rescue they estimate the value to be in the region of US$18 billion (€16.2 billion) per annum.

Benefits of improved weather forecasting to non-marine fields

The main benefits to other fields from enhanced ocean observation come through improved weather and climate forecasting. Some of the descriptive US case examples presented by Adams *et al* are identified below:

- Damages in the US from severe weather are nearly US$12 billion (€10.8 billion) annually (NOAA 2002). The proportion of the US GDP that is in one way or another sensitive to weather is estimated to be as much as one-third of the GDP or around US$3 trillion (€2.7 trillion).

- It has been estimated that ENSO (El Niño/Southern Oscillation) forecasts a year in advance with 60% accuracy could potentially save the agricultural, fisheries and forestry sectors of the US economy about US$500,000–1.1 billion (€450,000–1 billion) per event, or US$183 million per year (€165 million) over a 12 year period, based on a saving of US$732 million (€660 million) per event. This saving would increase to approximately US$300 million (€270 million) per year if the accuracy of ENSO forecasts improved to 77%, which is realistically achievable. California achieved US$1.1 billion (€1 billion) less damage in 1997–98 than it would otherwise have done, through warning of El Niño effects. The rest of the US saved US$200–300 million (€180–270 million) a year due to similar, smaller scale warnings.

- A study on the costs and benefits of ENSO forecasts concluded that, for agricultural benefits alone, the real internal rate of return for US federal investments in ocean observation for ENSO prediction is within the range of 13–25%. By incorporating ENSO forecasts into planting decisions, farmers in the US could increase agricultural output and produce benefits to the US economy of US$300–400 million (€270–360 million) per year.

Outside the US, ENSO studies have been performed in Australia, Peru, Brazil and Ethiopia. In Australia, every 0.5°C of ENSO-related cooling of the waters off northern Australia equates to about AU$1 billion (€540 million) in lost agricultural revenues (Hassall and Associates 1998). Three month seasonal outlooks have been issued for three years to advise the agricultural community on planting and harvesting decisions, and on the use of pasture lands. Experimental ENSO forecasts form the basis of an economically successful programme in Peru (soon to be introduced to Ecuador) that advises farmers on whether to plant rice or cotton (crops with very different water demands). In North-east Brazil, ENSO advisories have guided the planting of corn, rice and beans with encouraging results since 1988. In Ethiopia, advice based on ENSO forecasts has guided land use strategy, conservation policies and economic assistance policies.

Key findings

From the analysis of the costs of ocean monitoring and the related work on social and economic benefits, the list below identifies the key findings and the factors that influence ocean monitoring:

- A significant proportion of world economic activity and a wide range of services, amenities and social benefits depend on the wise use of the sea. The ocean plays a crucial role in sustaining life on Earth and is a key element in climate change. In the short term, better and more systematic observations of the ocean will enable us to forecast imminent disasters from storms, floods and drought, and mitigate their effects by warning the populations at risk.

- The current ocean observation system exists in scattered pieces. Observations are often taken from 'platforms of convenience', which may be ships at sea or old lighthouses. The former provide observations that are not sustained, while the latter take place at locations that may or may not be important from an oceanographic perspective. Data from many sources is rarely brought together and large expanses of the ocean remain unobserved for substantial periods of time.

- Activities are underway by 16 European partners, from 14 coastal states, to update their national European Directory of Marine Environmental Data (EDMED) entries and to develop and install an innovative infrastructure for updating the EDMED database by means of the Internet.

- The Sea-Search Network has been set up to provide an effective navigation tool to data and information sources in Europe and to centres in Europe with expertise and skills in oceanographic and marine data and information management.

- Particularly with regard to data and information systems, it would appear that there is considerable duplication of effort (real or apparent) at the international level and there may be confusion for potential users. A bewildering web of organisations and networks exists at the international level.

AIR QUALITY MONITORING

Cost of air pollution

An estimated three million people die each year because of air pollution (WHO 2000); this figure represents about 5% of the total 55 million deaths that occur annually in the world. It is possible, because of uncertainty in the estimates, that the actual death toll is anywhere between 1.4 and 6 million annually, which based on a 'willingness-to-pay' estimate equates to an economic evaluation of €1.26 trillion to €5 trillion per year.

Based on figures from a study in Austria, France and Switzerland (UNECE 1996), it can be estimated that costs for Europe (five million deaths per year) would be as much as €270 billion (mortality only), with morbidity representing a further 33% of the valuation, giving a total economic valuation of about €360 billion per year.

Economic valuation of the benefits

Using percentage values of 1%, 5% and 25% for Scenarios 1, 2 and 3 respectively, the economic valuation of the potential benefits of a GMES framework for air quality monitoring are €4 billion for Scenario 1, €18 billion for Scenario 2 and €90 billion for Scenario 3. These figures will depend on the strategic plans related to the benefit of saving lives and preventing health deterioration through the introduction of improved air quality policies and monitoring programmes. For example, the recent proposed amendments to the Commission Communication (EC 2003c) on the Clean Air for Europe (CAFE) programme have a number of paragraphs directly applicable to a GMES framework.

As regards the current air quality monitoring network in Europe, it is estimated that the total capital costs are around €30 million, with annual running costs of around €60 million (based on the UK's Central Management and Co-ordination Unit budget figures in 2002). Assuming integrated enhancements to the network, a saving of around 1% of the running costs might be achieved, amounting to €6 million per year.

Key findings

From the analysis of the costs of air quality monitoring and the related work on social and economic benefits, the list below identifies the key findings and the factors that influence air quality monitoring:

- Air pollution is a major environmental health problem affecting developed and developing counties alike. Air pollution damages plant and animal life and contaminates water sources, threatening economic and social welfare as well as health.

- Various EU directives and conventions exist of relevance to air quality. The number of air pollution monitoring sites in Europe is very large, but it is reported for several countries that the number of sites is being reduced as budgets become more tightly controlled. Gaps in the national monitoring networks are most obvious in the accession countries. The CAFE programme aims to strengthen air pollution policy, based on the best available science, and to enable a broad, open and transparent dialogue with the scientific community, as well as the public and the stakeholders. Airbase, the air quality information system of the European Environment Agency (EEA), contains a database carrying information submitted by participating countries from across Europe. However, this database is not easy to use and it would appear that much of the European level air quality data is out-of-date in comparison to national level data in many European countries.

- A number of organisations and networks exist at the international level (or European regional reporting level to international organisations) such as the World Meteorological Organization (WMO), the World Health Organization (WHO) and the United Nations. There are a number of important international and European programmes, including the Convention on Long-Range Transboundary Air Pollution (CLRTAP) and the European Monitoring and Evaluation Programme.

CLIMATE CHANGE RESEARCH

Benefits of climate change research

As outlined in the Intergovernmental Panel on Climate Change reports (IPPC 2001), sustainable development in the past in Europe has been in jeopardy from several existing pressures, mostly non-climatic (for example, land use change, environmental pollution, atmospheric deposition), yet climate change adds an important element to the threat to the environment. Sea level rise threatens coastal habitats with a squeeze between hard defences and rising water levels. Most (50–90%) of the European alpine glaciers could disappear by the end of the 21st century, and there may be local extinctions of species that require cold habitats for their survival (for example, sub-arctic and montane species). Many ecosystems will respond to climate change via migration and change; a policy challenge is how to manage these changes.

Climate change impacts will be differently distributed among different regions, generations, age classes, income groups, occupations and genders. This has important equity implications, although these implications have not been investigated in detail. For example, elderly and sick people suffer more in heat waves. There is greater vulnerability, in general, in southern Europe than in northern regions. Mediterranean and mountain farmers are likely to be worse off in a warmer world. This presents a challenge to existing regional policies within the EU that are aimed at levelling up less developed areas. In general, the more marginal and less wealthy areas will be less able to adapt, so climate change without appropriate policies of response may lead to greater inequity.

Possible climate change impacts on key resources are sufficient to warrant early consideration by European policy makers to ensure sustainable development. In general, the adaptation potential of socio-economic systems in much of Europe is high because of economic conditions, a stable population with the capacity to move within the region, and well-developed political, institutional, and technological support systems.

Economic valuation of the benefits

Peck and Teisberg (1993) calculate a present value of US$50 billion (€45 billion) for resolving specific uncertainties about climate change now rather than in 40 years time. The US leads the world in research on climate and other global environmental changes, spending approximately US$1.7 billion (€1.53 billion) annually on its focused climate change research programmes (IGFA 2002). This contribution is roughly half of the world's climate change research expenditures, three times more than the next largest contributor, and larger than the combined contributions of Japan and all 15 nations (in 2003) of the EU. In total, the European budget is around US$1.1 billion (€1 billion). The high economic demands and scientific complexity of climate change research push the field towards being more applicable to European level implementation (without compromising national leadership and autonomy) and the sharing of facilities and expertise in order to make the most of European investment.

Key findings

From the analysis of the costs of climate change research and the related work on social and economic benefits, the list below identifies the key findings and the factors that influence climate change issues. The most significant impacts of climate change in Europe that will require greatest attention with respect to policies of response will be as follows:

- The high vulnerability to climate change in the south of Europe and in the European Arctic.

- The risk of water shortages in southern Europe.

- Flood hazard is likely to increase across much of Europe.

- The risk of fires will increase in southern Europe.

- Some agricultural production systems in southern Europe may be threatened by increasing drought conditions.

- The insurance industry faces potentially costly climate change impacts through property damage, but there is great scope for adaptive measures if initiatives are taken soon.

- Developments on the coasts will be exposed to sea level rise and extreme events, necessitating protection or removal.

The benefit of GMES to climate change research could be at least €3.5 billion per annum in the early stages of GMES operations, but would be anticipated to rise considerably if GMES performs a prediction or forecasting role. Climate change has the most to gain from combining research capacities at the European level and in return the GMES programme (and European citizens) has the most to gain from the successful implementation of such an integrated, multi-disciplinary, multi-player theme. Initiatives at European level include the European Climate Support Network, the European Climate Forum and the European Climate Computer Network.

OTHER QUANTIFIED BENEFITS

During the course of preparing information on the case studies, a number of other valuations have been calculated and these provide some initial or example valuations. The various categories of GMES added value are outlined below.

Improved efficiency of information production

It is anticipated that GMES could play an important role in improving project implementation through the utilisation of common datasets and tools, access to information, etc, in ensuring that important project results are not lost and in avoiding the duplication of previous efforts across the various DGs and framework programmes. Based on just the number of European Commission projects recorded in its CORDIS information system data for natural hazards, ocean monitoring, air quality monitoring, air quality health projects and climate change research, the savings could be as much as €56 million per annum.

Avoided duplications

Before the existence of EuroGOOS, each of nine European countries would map the whole Arctic Ocean. Within a GMES framework it should be possible to share data and avoid such duplications. Based on scaled figures from the US Global Ocean Mapping Project (Vogt *et al* 2003), the total saving would be around €4.6 million.

The INSPIRE Framework Definition Support (FDS) working group estimated that savings of at least €250 million per year could be saved by avoiding duplication in the development of spatial datasets across the EU25 countries (EC 2003b).

Sharing common parts of observing systems

Before the adoption of the new EU regulation concerning monitoring of forests and environmental interactions (No (EC) 2152/2003 of 17 November 2003), observing systems for forest fire and forest acidification used to be totally separated. Under the new framework, such observations will be made in a more coherent way. Based on three man-years effort across Europe for each pair of common parameters, the saving would be around €300,000.

Co-funding and re-use of datasets of common interest

Satellite scenes used to be acquired on a case-by-case basis by a variety of European Commission and Member State services. As part of the CORINE (Co-ordination of Information on the Environment) land cover inventory of the EEA, the Joint Research Centre (JRC) has created a Europe-wide archive of ortho-rectified Landsat 7 scenes for the year 2000, which is accessible free of charge. The budget for this activity was around €765,000. Based on an anticipated 4:1 cost benefit ratio (or return on investment) for the re-use of common or shared datasets (Korte 1996), the net benefit should be a minimum of €2.3 million for the exercise to be deemed beneficial.

Ensuring data compatibility

Incompatibilities between data from different sources (for example, land use statistics and land cover geographic data) reduce information quality and increase costs. The INSPIRE FDS working group estimated that savings of at least €800 million per year could be saved by efficiency gains in European environmental impact assessments and strategic impact assessments, environmental reporting and protection, implementation of environmental *acquis* and expenditure on trans-European transport networks.

Improved access to data and information

The European Soil Map was created by the European Commission. However, copyright restrictions limit its re-use by other teams. The cost of this activity is estimated to be around €500,000. Again based on a 4:1 ratio for this type of activity, the net value of the benefit of multiple re-use of the datasets should be at least €1.5 million.

Autonomous information

CORINE land cover maps for the whole of Europe at 1:100,000 scale were produced for 1990 and 2000 base years (at a project cost of around €10 million) from Landsat data. A risk exists that the Landsat series is interrupted. If GMES secures operational satellites, the benefit would be of a strategic nature. Instead of having to depend upon US data, the existence of up-to-date independent information on land cover would help secure subsequent inventories servicing EU policies on environment, security, agriculture and regional development.

PROCESSES FOR PRIORITISATION

To a certain extent, this socio-economic analysis can be used to initiate prioritisation, and indeed the Extended Impact Assessment methodology includes a mechanism to undertake a comprehensive process of prioritisation. The case studies and topics discussed in this chapter with the highest economic and social importance in Europe are (in priority order): air quality, flooding, earthquakes, windstorms, droughts. Return on investment (ROI) ratios can be used to determine how much of the potential benefits of improved monitoring in a particular theme can be expected to be met by set levels of GMES investment. A quick summary of the selected case studies would indicate Scenario 1 type annual benefits based on economic evaluations only in 2009–10 (at 2003 figures) as follows:

Earthquake monitoring	€1 billion
Volcanic monitoring	€2.5 billion
Ocean monitoring	€4 billion
Air quality monitoring	€5.5 billion
Climate change research	€3.5 billion

Ratios for ROI could be set between 2:1 and 7:1 (Korte 1996), realising that this can and will change over time. As an example, based on a ROI ratio of 3:1, a GMES annual budget of around €5 billion would be required to address the full potential benefits of improved monitoring of the themes outlined above. Using a simple scaling of high, medium and low potential benefits for monitoring of a total of, say, 12 environmental themes, this could escalate the required annual GMES budget to around €12–15 billion. Of course, on lower budgets, while retaining reasonable ROIs, GMES could address a broad range of themes by not proposing to approach such high levels of improved monitoring; however, this runs the risk of perceived failure to improve environmental monitoring in any one theme.

CHAPTER 12

CONCLUSIONS AND RECOMMENDATIONS

MAIN DATA POLICY CONCERNS

The purpose of this chapter is to draw together the major common issues in this book and to propose recommendations for future action on how to improve data access for global monitoring. The first part of the chapter identifies the main concerns on data policy and access to data that have been common themes throughout the book. The question of sustainable funding is one of the cornerstones of building environmental data systems, so this issue is examined separately. The chapter is completed with a discussion of recommendations and next steps.

During 2003, a set of five working groups drawn from the European Commission and European Space Agency (ESA) Member States discussed subjects related to Global Monitoring for Environment and Security (GMES) and global environmental monitoring. One of these working groups was concerned with data policy. During their discussions, the members of the working group identified the following six points that contribute to answering the question of what is the purpose of a data policy for GMES, or more widely for global environmental monitoring:

(1) To promote the use of services, information and data in order to maintain/reach leadership in spatial data and related technologies.

(2) To promote collaborative and multiple use of services, information and data.

(3) To take into account the existing and emerging data policies of the main actors such as ESA, the EU, national institutions and commercial providers.

(4) To promote European business in order to maximise commercial investment and to attract private funding.

(5) To promote the availability of convenient and consistent standards, calibration and metadata, also including aspects such as rectification, calibration, atmospheric correction, accuracy assessments, product advice and specifications of the analysis process.

(6) To ensure long-term archiving, particularly in the case of commercial data suppliers as the market value of data often falls with age.

Taking into account these six points, plus the issues and concerns that have emerged on data policy in the earlier chapters, this section identifies the main data policy issues and challenges. This section synthesises the earlier material and presents 10 key concerns that will have to be addressed in the development of data policy in the GMES context. For convenience the issues are given in alphabetical order, deliberately to avoid an impression of priority.

Archives. These are increasingly in digital form, although many datasets/ archives relevant to environmental policy are in hard copy form. Digital data archives have presented physical problems in media and reading machines, but there are even greater challenges of storing and accessing hard copy archives. Some organisations have scanned documents and aerial photographs in their possession, for example, the World Health Organization (WHO) and some European national mapping agencies. The mix of private sector and public sector organisations in the environment sector will present challenges because the motivations of these two types of organisation are different in terms of data archiving. The private sector has an interest in sales of data and products and so datasets that do not sell can be regarded as a burden on a business, while the public sector has some form of responsibility for safeguarding resources that can be beneficial to the public in the long term.

European spatial data infrastructure. While Europe has many environmental services, for example the sustainable development indicators of Eurostat or the Joint Research Centre (JRC) soil bureau, there is a view that the quality of the products provided by these services could be substantially improved by developing a European spatial data infrastructure. Europe is falling behind in this respect. The Infrastructure for Spatial Information in Europe (INSPIRE) initiative has been concerned with infrastructure for spatial information in Europe and has focused on improved access to information related to the environment in Europe.

Internet. The Internet is changing pricing and distribution policy. In many institutions (for example, for population censuses) there is a shift from providing data at the cost of reproduction to providing data free of charge. This reflects the logic of the short-run marginal cost (Harris 1997). When it is necessary to produce and deliver a CD to hold a dataset there is a cost of materials, labour, packing and posting. When the dataset is made available on a server there is no cost to the data provider in delivering the data, so the cost of reproduction falls to zero. The information society thrives on the Internet, but variable access to the Internet is contributing to a digital divide in environmental data.

Legal obligations. Legal obligations are important in driving data accessibility. Examples of legal obligations include: (1) the German law on environmental information (*Umweltinformationsgesetz*), which states that all data concerning nature and environment in public authorities in Germany must be freely available and free of charge; (2) the European Community Directive on Freedom of Access to Environmental Information; and (3) the provision of national statistics to Eurostat. The growth in formal, legal obligations on government departments is often driving data accessibility conditions.

Licences. It is common for licences for the use of environmental data to be restricted to single projects or single applications. Given that environmental data is by its nature of wide application, there is a demand by users to have more flexible licence arrangements to enable and to encourage wide use of data.

Map access. Mapping agencies tend to use national standards rather than international standards. This presents challenges for Europe-wide map datasets. For many global environmental monitoring applications, specifically on topics related to regional development and humanitarian aid, access to map data will be required. In many parts of the world, maps are only openly available at a best map scale of only 1:200,000 (for example, Russia) or even 1:1,000,000 (for example, China), or not at all.

Pricing. Approaches to pricing are important signals to the market for environmental information. A common model used by environmental information providers is to set a market price, that is, the price that the market will bear (which may be zero in some applications). For activities that contribute to the public good, for example scientific research, environmental data are often free or made available at the cost of reproduction. The challenge here lies in the transition from research to operational systems. Mature operational systems that use environmental data (for example, in meteorology) can readily justify the costs. Projects that show operational potential find it more difficult to justify the costs of environmental data because the benefits are not easy to quantify or to capture only by predictions.

Privacy and confidentiality. Some environmental data collected in Europe is not available to a wider audience because of commercial sensitivity (for example, the location of oil spills), ecological protection (for example, nesting sites of endangered species) or national security (for example, locations of nuclear installations). This may mean that some datasets are not available for European environmental and security applications, and therefore only a partial coverage is possible.

Public good. The concept of public good is common in many organisations that provide environmental data (Pearce 1995). It is typical for there to be few restrictions on dissemination of data if the application or use of the information is for the public good, while there is typically control of the data if provided for commercial applications. It is essential that the case for a public good is considered in parallel with the need to ensure a sustainable funding base and to enable the continued routine production of information for environmental policies.

Standards and metadata. These continue to present problems because many different systems are used and convergence is limited. The best practice is in meteorology, which is characterised by being operational and international. An open question here is how can community consensus on streamlined standards and metadata be stimulated and then be effective? The work of the International Standards Organisation, for example ISO 19115 on environmental data, and the work of the Open Geographic Information System (GIS) Consortium are increasingly important and relevant to environmental monitoring. The community of users and of data and product suppliers need to see real benefits from the adoption of standards in order to justify the resources needed to ensure that standards are applied.

SUSTAINABLE FUNDING: CHALLENGES AND OPTIONS

Introduction

One of the main challenges for GMES and other global environmental monitoring programmes lies in securing sustainable funding. While GMES is, at its origin, an initiative concerned with European independence and autonomy in environment monitoring and security, the justification of sustainable funding is a necessary condition for developing an operational GMES. This section discusses the main dimensions of the challenge of sustainable funding.

Public and private sector funding

There is a difference between funding from the public sector and from the private sector. Funding from the private sector has a commercial justification and ultimately is provided to increase shareholder value in the companies involved. Public sector funding on the other hand is provided for public objectives such as meeting public policy requirements, improving the quality of life of citizens or in the provision of public health services.

A distinction can also be made between sponsorship funding and customer funding. The distinction is between the funding being available as a stimulus to activity, such as research and development, or the funding being used by a customer to purchase goods or services to meet their own needs. Given the broadly public good nature of environment and security, it is likely that the customers will be in the public sector. Note the difference here between public sector funding *per se* and public sector funding as either sponsor or customer. Under the GMES approach, public sector funding in sponsorship mode will be provided under two circumstances:

(1) to develop the infrastructure components of GMES; and

(2) to construct and operate GMES for European strategic reasons if operational customers (public or private) are not willing to pay for goods and services through GMES.

Public and private goods

A public good is not a good provided to the public by a public authority. A public good has two main characteristics (Pearce 1995): non-rivalry and non-excludibility. Non-rivalry means that the use of the information does not diminish the capability of another user to use the information. Non-excludibility means that no one user can be excluded from using the information by another user. The classic example of a public good is a lighthouse. A lighthouse provides light to all users under the non-rivalry and non-excludibility principles. However, some lighthouses are built and operated on a private basis (through charities, for example) and some by public bodies (regional authorities, for example).

A private good, on the other hand, is a good or service whose consumption by one person or organisation excludes consumption by others.

Geospatial data display some of the characteristics of a public good (Ordnance Survey 1996). Now that much geospatial data is in electronic form, the potential public good nature of the data has increased: there is no rivalry between users as each user can obtain a copy of the data with no deterioration in the quality of the data, and one user does not necessarily exclude another user from using the same information.

The concept of a hybrid good may also be useful. Some government departments buy data and products for their own purposes, which ultimately have a public good basis, for example predictions of extreme environmental events. Once the data has been used for its original purpose, they can be made available for other applications if they are useful. The UK funding for the Advanced Along Track Scanning Radiometer (AATSR) on Envisat is a good example of this. The UK Department of the Environment (now the Department for Environment, Food and Rural Affairs) provided funding of approximately £10 million (€14 million) to build the AATSR instrument. The reason for this funding was to continue the measurements of sea surface temperature begun by the Along Track Scanning Radiometer (ATSR) instrument on ERS-1 (European Remote Sensing satellite) and continued on ERS-2. The Department of the Environment needed these measurements in order to fulfil its own policy objectives of identifying the human-induced fingerprint of climate change. The public good element here is the provision of policy advice on climate change. Once the AATSR data has been used for its primary public good purpose of sea surface temperature measurement in support of policy on climate change, the data is made more widely available for other applications that may be private sector (or private good) in nature. Therefore, the data is used for both a public good purpose and a private good purpose, a hybrid of the two.

Economic theory

The challenge of sustainable funding always involves views and ultimately policies on pricing for data and for information. The basis for cost-recovery can be assisted by economic theory. A number of market and pricing conditions can be envisaged for all information (Ordnance Survey 1996), whether environmental or not as follows:

- A competitive market in which there are a large number of well-informed buyers and sellers. A stock market fits this description.

- Classic monopoly with only one supplier.

- Natural monopoly where the very high fixed costs and low marginal costs make it inefficient for more than one firm to supply the market because of the duplication in fixed costs involved.

- Price differentiating monopolist where the fixed costs of supply are recovered from higher value users, but users with a lower willingness or ability to pay are charged a lower price.

- Dissemination cost pricing where the price is set at the marginal cost of dissemination. Where dissemination cost pricing is supported by government subsidy this results in a sub-optimal allocation of resources since the government subsidy is normally greater than the benefit to users of the lower price achieved by dissemination cost pricing.

Pricing models

These theoretical options can be seen in practice in the pricing models used for environmental data. Environmental data is typically made available under a variety of pricing models, namely free data, marginal cost price, realisable price, full cost price, two tier price and information content price. This section comments on the implications of each of these pricing models for the sustainable funding of global monitoring. Harris (2002) gives for each pricing model the arguments in favour and the arguments against, while this section concentrates on the implications for sustainable funding.

Free data for all users. Providing data free to all users maintains the cost of providing the data on the data supplier. Sustainable funding can only be guaranteed therefore when the data supplier has a secure and continued source of funding for data supply. This implies a public source of funding, typically in support of a government policy requirement.

Marginal cost price to all users. This price recovers the marginal cost of providing data to a user, although for data provided over the Internet this cost is likely to be zero as the data transmission cost is not met by the supplier. Where there are actual costs (for example, media and postage), then the supply of data is charged at these marginal cost rates. The case for sustainability is the same as for free data.

Market driven or realisable price for all users. All users pay a price that the market will bear. This is broadly the case at present for commercial Earth observation data provision. From a supplier perspective the market price would reflect the full economic cost of carrying out the data collection and supply activities, plus a profit in the case of commercial activities. Market prices could be sustainable as long as the market is mature and there are willing buyers who are prepared to pay prices that enable suppliers to cover their costs and make a profit. Market prices are not sustainable where the price is below the cost-recovery level, not least because this does not allow investment in new data collection activities.

Full cost price. The full price covers all investment costs plus a profit. In a commercial sense this is a sustainable model as long as the data and information buyer market is sufficiently large.

Two-tier pricing. The two-tier pricing model typically consists of: (1) a marginal cost price (or free) for research users operating (in a broad sense) for the public good, and (2) all other users who pay a market price or a full cost price. This model is sustainable as long as the open market is sufficiently large to allow research activities to benefit from the core, sustainable funding.

Information content pricing. Users normally want information and not data, so pricing by information content (for example, the location of an oil slick or a diseased crop) is ultimately more logical than pricing by data volume. However, the debate shifts to the value of the information, which in turn relates to the role of the information within each buyer organisation. Where organisations can attribute value to information clearly, then this model could contribute to sustainable funding.

Closed and open loops

The debate on pricing is often expressed as whether the user should pay or the supplier should pay. The nature of the system (either closed loop or open loop) will have significance for the debate on pricing policy and sustainable funding. For closed loop systems the value of the data is defined before the investment is agreed. In open loop systems the investment decision is made on the basis of anticipated value. Examples of closed and open loop structures are presented in Figure 12.1.

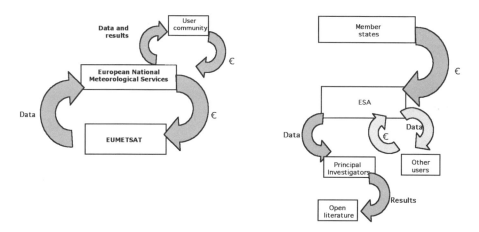

Figure 12.1: Examples of closed loop (left side) and open loop (right side) funding systems in EUMETSAT and ESA respectively

A closed loop system is operated by EUMETSAT. EUMETSAT is funded by its European National Meteorological Service (NMS) partners, which receive and then use the data for their own internal purposes, and can also distribute data and products to other users within their own territories at a charge to those users.

An open loop system is operated by ESA. The ESA Member States provide funds for ESA which then provides data to principal investigators and pilot projects, and also sells data to other users through the distributing entities: the principal investigator or other user may or may not be linked to the organisation within a Member State that provides the initial funding.

In a closed loop system, the primary user pays for the whole data capture system. There is no charge for data at the point of delivery to the user because the primary user has already paid for the system. In a closed loop system, the value of the investment is therefore assessed by the users who continue to pay for the system based upon the benefits of the investment.

In the case of an open loop system, the situation is characterised by the supplier paying a relatively large proportion of the cost of providing data because the price to the user of data and products is typically relatively low.

To establish and operate a European capacity for global environmental monitoring, the preferred funding approach in the long-term is a closed loop where the user has a clear requirement for information that is satisfied by data and information. A good case in point can be seen with government departments in Europe that have a need for information on climate to support their climate change policy responsibility. Such departments could act in a closed loop manner to fund data and product acquisition to meet their own needs.

However, it should also be recognised that sources of open loop funding are indispensable as part of the development phases of GMES, and for some of the pre-operational and operational services. The funding provided in 2003 and 2004 by the European Commission for the GMES Thematic Projects and by ESA for the GMES Service Element projects are examples of open loop funding. One objective of the funding for these projects by the European Commission and ESA is to contribute to an eventual sustainable funding base for applications within GMES.

RECOMMENDATIONS

Introduction

This chapter now turns attention to what might be done in GMES on access to data. The lessons are also valid to other developments in global monitoring. The recommendations below indicate the main issues that need attention. The discussion of recommendations is divided into the following characteristics:

- control of data and information;

- technical accessibility;

- quality approval;

- costs and funding;

- archiving; and

- European spatial data infrastructure principles.

Control of data and information

New technologies are developing sufficiently quickly to allow transmission of all the data available within a European capacity for global environmental monitoring. However, at face value this could mean a loss of control over the data. A system of encryption and decryption to control access to environment and security data and products could be implemented. All real time data could be transmitted to all users in a broadcast mode, but in an encrypted form. Access to the true data could then be controlled by decryption keys, which could be made specific by a number of criteria including space, time, product and use or user category.

The decryption strategy can easily take account of the type of use of the data. For example, some global monitoring datasets could always be provided free of charge, while all relevant data provided for humanitarian aid could be made available for free only at the time of a crisis.

EUMETSAT already uses encryption/decryption to control access to Meteosat products. The six-hourly Meteosat High Resolution Imagery (HRI) transmissions (that is, images disseminated at 0000, 0600, 1200 and 1800 hours each day) are unencrypted and available free of charge. HRI transmissions disseminated at half-hourly, one-hourly or three-hourly intervals are encrypted, and access to this data requires the signing of an HRI User Agreement with EUMETSAT, a decryption unit and a Meteosat Key Unit.

Another example of the use of encryption/decryption is the developing European Galileo positioning system. The Galileo system is planning for the following four levels of service access, controlled by encryption/decryption of the signal transmitted from the Galileo satellites (von der Dunk 2003):

(1) Open service. Free, openly available, no charge, but no integrity and no liability.

(2) Commercial service. Available at a fee, with a service guarantee and liability by the supplier. Encrypted.

(3) Safety-of-life. High integrity and high availability. Encrypted.

(4) Public-regulated service. Secured, closed access. Encrypted.

The Internet offers the most suitable standard option for the transmission of encrypted environmental data. The Internet is widely used as the main means of data dissemination by relevant organisations already, for example many of the statistical institutes in Europe. For high volume data transmission, the current developments in e-Grid technology provide even greater opportunities for rapid data dissemination.

One foreseeable difficulty is the use of the Internet in some parts of the Less Economically Developed Countries (LEDCs), in particular in sub-Saharan Africa. Some regions in LEDCs have only limited telephone connectivity and therefore limited Internet access, and yet these areas could substantially benefit from operational environmental monitoring, in particular as part of humanitarian aid and regional development activities.

Most Internet applications use land-line technology, but satellite data dissemination provides opportunities for data distribution. A number of Digital Video Broadcast (DVB) satellite systems support high speed Internet. Data transmission speeds of 30 Mbps are available at present from satellite DVB, a speed that is not incompatible with the 100 Mbps downlink speed typical of satellite radar data, for example. Statistical and map data would present lower speed requirements. Such DVB systems may provide data dissemination to locations within the downlink coverage of the DVB satellite. This would be comparable to disseminating environmental data and products in a manner similar to satellite television channels. The case of encryption/decryption is particularly interesting here as this is exactly how satellite television operators control access to their broadcast programmes.

Technical accessibility

The problems over metadata standards are twofold. First, there are many standards in use. Secondly, many users ignore any standards at all. Without the use of agreed standards by all participants, any environmental data system or service infrastructure will be reduced to a data depository. There are at least two tracks in adopting and encouraging the use of standards:

(1) the developments in metadata standards in the International Standards Organisation and in the Open GIS Consortium; and

(2) market standards, such as Extensible Markup Language (XML) and those used commonly in GIS.

The environmental community should push for the adoption of widely used standards, including market standards, rather than specific approaches that might be technically appropriate but are not widely employed by users.

Quality approval

The control of data dissemination can be achieved through the technical means of encryption/decryption, as described earlier. Technical dissemination is also closely linked with copyright, intellectual property rights (IPR) and licensing, because these characteristics provide the legal framework for data protection, dissemination and use.

Copyright and IPR should not be seen solely as ways of controlling (that is, restricting) access to data and information, but as a positive means of declaring the quality of environmental data and products. The Canadian government, for example, views copyright as a symbol of quality in data rather than as a way to restrict data use. Licensing agreements can be used to document the rights and responsibilities of supplier and user, to protect the quality of products, and to increase recognition and branding of good quality environmental data and products.

The use of copyright, IPR and licensing to improve quality and branding is even more valid for scientific model outputs. The suppliers of model outputs have a responsibility to ensure the quality of their products. For example, the model outputs that are generated by the European Commission Thematic Projects and the ESA GMES Service Element projects (and their follow-on actions) could benefit from having a quality approval that is defined in terms of copyright, IPR or licensing. In this context, the use of agreed standards for metadata is a clear example of formal responsibilities of the model output providers.

Costs and funding

Data pricing policies should allow for flexibility: data for humanitarian aid purposes may be made available free of charge at the time of the emergency and for authorised users, but the same data may carry a charge for other uses. This could allow, for example, the inclusion of very high spatial resolution data both serving a humanitarian aid purpose and fulfilling the commercial objectives of the data supplier. A similar argument applies to value added services. Information products that normally carry a charge may at times be made available for free by the value added product supplier on their decision, for example in cases of severe pollution events. The encryption/decryption approach noted above would provide the technical means of allowing control of the data and information products. Several different pricing models could be used, tailored to different circumstances.

A common complaint from users is that datasets acquired for one purpose by an organisation cannot be used for another purpose or on another project. Environmental monitoring programmes could usefully negotiate licences with data suppliers where prices are agreed for wide access by all approved participants in a programme. This would need a negotiated balance between (1) the income that a supplier would lose by not having multiple sales of the same data against (2) the higher price for a licence that allows a larger number of users to access the data. Access to data by larger numbers of users is likely to increase the use of environmental data and products anyway and so expand the market for environmental information to the potential benefit of both the suppliers and the users.

Open loop funding is entirely appropriate for research and development activities, and funding is often provided in sponsorship mode. To achieve sustainable funding in the longer term it is desirable to secure closed loop funding where there is a user or customer for environmental data and products. The customer should fund such information provision to satisfy its own needs and, as long as those needs exist, will provide sustainable funding. An important challenge before the full development of operational systems is in the transition from sponsorship funding to customer funding where the user pays. There are already existing funding sources that operate in this way, for example from national administrations for the monitoring of air quality, from EU policy programmes (for example, the EU Natural Disasters Fund) and even from private sources such as companies responsible for industrial plant emissions. As an example, Figure 12.2 overleaf shows a closed loop funding example where a government department or agency buys data and information to contribute to its climate change policy. In addition, the government department or agency allows the supplier to sell its data and information to the wider community under controlled conditions. The control can be achieved legally through IPR and technically through encryption and decryption.

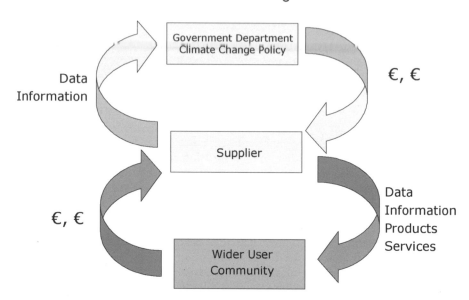

Figure 12.2: An example of closed loops to support the climate change policy needs of a government department or agency. The data and information can then be made more widely available through controlled dissemination

Archiving

There are two main problems with current environmental data archives. First, if the missions or data collection systems are commercial, as the data ages there is a reducing likelihood of sales of the data and an associated reduction in incentives for preserving data. Secondly, in the public sector-funding for archives is typically not long-term in nature, but tends to be on a project or mission basis. In order to contribute to securing a commitment to the long-term safeguarding of environmental data, we recommend that the European Commission establish an appropriate legal instrument to ensure that if substantial environmental datasets are to be destroyed by European organisations (public or private) then a public body is given the right to accept the data if it wishes. This recommendation would result in a situation in Europe comparable to the US National Satellite Land Remote Sensing Data Archive and to Archives New Zealand.

European spatial data infrastructure

The experience gained in the context of the INSPIRE initiative on spatial data can usefully be extrapolated to the wider remit of environment and security information. The five main principles debated within INSPIRE (see Chapter 9 of this book) could be adopted by GMES as a useful starting point. During the INSPIRE consultation phase, 85% of respondents agreed with the need to establish a common data policy framework to share datasets between public bodies in Europe.

CHALLENGE

The initiative on GMES is a bold and some would say brave development in Europe. While GMES is specifically European, there are many characteristics in common with any global monitoring programme. There is no doubt that political will is needed to leverage the relatively large financial investments envisaged. Even when finance is justified and committed, because global monitoring covers a wide range of disparate datasets there will always be the challenge of enabling access to data when the organisations providing the data have different objectives, agendas and modes of operation.

In February 2004 the European Commission (working closely with the ESA) presented a Communication to the European Parliament and the Council (European Commission 2004a). The Commission reported on the work so far in GMES and indicated the direction of future activities. Given the broadly socio-political nature of GMES, the Communication may be an important step on the road to funding of the order of hundreds of millions of euros for global monitoring by European institutions. However, the Communication does emphasise the challenge of access to data in global monitoring and states right at the start:

> At a time when command of information has geo-strategic implications, investments have been and continue to be made at various levels, *without co-ordination* [emphasis added], to develop monitoring technologies and data systems.

In the rush to build infrastructures, the challenges of access to data must not be forgotten.

Organisations from which information was collected

Airborne Remote Sensing Facility, UK
Australia and New Zealand Land Information Council
Australian Bureau of Meteorology
Australian Bureau of Statistics
Australian Surveying and Land Information Group
Austrian Space Agency
British Atmospheric Data Centre
British National Space Centre
Bundesamt für Naturschutz, Germany
Centraal Bureau Voor de Statistiek, The Netherlands
Central Statistics Office, Ireland
Commonwealth Bureau of Meteorology, Australia
Danish Environmental Protection Agency
Danish Forest and Nature Agency
Danish Meteorological Institute
Danish Ministry of Environment and Energy
Denmark Statistik
DigitalGlobe, US
Earthwatch, US
Environment Canada
EUMETSAT
EuroGeographics
EuroGeoSurveys
European Centre for Medium-Range Weather Forecasts
European Centre for Nature Conservation
European Commission
European Environment Agency
European Space Agency
European Union Satellite Centre
European-Mediterranean Seismological Centre
Eurostat
Finnish Environmental Institute
Forestry Commission, UK
GeoConnections, Canada
GetMapping plc, UK
Institut Geographique National, France
Institute for World Forestry
Instituto Nacional de Estatistica, Spain
Instituto Nazionale di Statistica, Italy
Instituto Portugues de Cartografia e Cadastro, Portugal
Irish Meteorological Service
Japan Map Centre

Joint Research Centre
Koninklijk Nederlands Meteorologisch Instituut
Kort-og Matrikelstyrelsen, Denmark
Lantmateriverket, Sweden
Medical Research Council, UK
Meteo France
Meteorological Office, UK
Ministerio de Medio Ambiente, Spain
Ministry of the Environment, Spain
National Aeronautics and Space Administration, US
National Council for Forest Research and Development, Ireland
National Environmental Research Institute, Denmark
National Institute for Coastal and Marine Management, The Netherlands
National Land Survey of Finland
National Oceanic and Atmospheric Administration, US
National Statistics, UK
National Survey and Cadastre, Denmark
Natural Environment Research Council, UK
Natural Resources Canada
Naturvardsverket, Sweden
Norwegian Meteorological Agency
Ondersteunend Centrum GIS-Vlaanderen, Belgium
Ordnance Survey of Ireland
Ordnance Survey, UK
RADARSAT International
Scottish Environmental Protection Agency
Sección de Información Meteorológica, Spain
SkogForsk, Sweden
Space Imaging, US
State Institute of Statistics, Turkey
Statens Foruenningstilsyn, Norway
Statisches Bundesamt Deutschland
Statistics Canada
Statistics Norway
Statistik Austria
Statistik Belgium
Statistiska Centralbyran, Sweden
Swedish Meteorological and Hydrological Office
Topografische Dienst, The Netherlands
Umweltbundesampt, Austria
United States Geological Survey
US Environmental Protection Agency
US Natural Resources Conservation Service
World Health Organization
World Meteorological Organization
World Trade Organization

ADDITIONAL SOURCES OF INFORMATION ON SOCIO-ECONOMIC BENEFITS OF GLOBAL MONITORING

Geohazards monitoring

Burgmann, R, Rosen, P and Fielding, E (2000) 'Synthetic aperture radar interferometry to measure Earth's surface topography and its deformation', *A Rev Earth Planet Sci* 28, 169–209

EU (2002) 'First meeting on the preparation of a Communication on an integrated EU strategy on prevention, preparedness and response to natural, man-made and other risks – Brussels, 11 December 2002', available at http://europa.eu.int/comm/environment/civil/prote/cp_integrated_en.htm

EU (2004) *The Planet Under Pressure,* available at http://europa.eu.int/comm/research/leaflets/changes/en/index.html

EU (2004) *The Threat of Natural Disasters*, available at http://europa.eu.int/comm/research/leaflets/disasters/en/index.html

IGOS Geohazard Theme (2002) *IGOS-GeoHazards Theme Proposal Version 4 (17/05/2002),* available at http://dup.esrin.esa.it/igos-geohazards/pdf/IGOSThemeProposal.pdf

Massonnet, D and Feigl, KL (1998) 'Radar interferometry and its application to changes in the Earth's surface', *Rev Geophysics* 36, 441–500

United Nations (2002) *Living with Risk: A Global Review of Disaster Reduction Initiatives,* available at www.unisdr.org/eng/about_isdr/bd-lwr-2004-eng.htm

US Geological Survey (1998) *An Assessment of Seismic Monitoring in the United States: Requirement for an Advanced National Seismic System. US Geological Survey Circular 1188*, Boulder, CO: US Department of the Interior

Vetere, AL and Kelman, I (2003) 'The EU role in risk and disaster management', *CURBE Fact Sheet 8*, available at www.arct.cam.ac.uk/curbe/FS8EURoles.rtf

Oil spill detection and ocean monitoring

Brown, M (1997) 'Cost/benefit analysis of GOOS – some methodological issues', in Stel, J (ed), *Operational Oceanography: The Challenge for European Co-operation, Proceedings of the First EuroGOOS Conference*, The Hague, October 1996, Elsevier Oceanography Series, 62, pp 286–93

Brown, M (2000) 'Valuing marine activities in Europe: provisional estimates, concepts and data sources', *Proceedings of Second EuroGOOS Conference*, Rome, March, 1999

Castelucci, L (2000) 'Economic assessment of the value of marine industries and services and user requirements', *Proceedings of Second EuroGOOS Conference*, Rome, March 1999

Chen, C and McCarl, B (2000) *The Value of ENSO Information to Agriculture*, Department of Economics, Texas A&M University

Costanza, R (1999) 'The ecological, economic, and social importance of the oceans', *Ecological Economics*, 100, 212

Exxon Valdez Oil Spill Trustee Council (2004) *Oil Spill Facts*, available at www.evostc.state.ak.us/facts

Flemming, NC (1999) *Dividends from Investing in Ocean Observations: A European Perspective*, paper presented at the OceanObs Conference, CNES, St Raphael, October 1999

GCOS Working Group on Socio-Economic Benefits (1995) *The Socio-Economic Benefits of Climate Forecasts: Literature Review and Recommendations*, available at www.wmo.ch/web/gcos/Publications/gcos-12.pdf

Huxley, G (1990) 'A calculation of the contribution of marine industries to the gross national product of the United Kingdom', in *Marine Technology in the United Kingdom, Committee on Marine Science and Technology, Annexe 19*

IACMST (2001) 'Climate of UK Waters at the millennium – status and trends', *IACMST Information Document No 9*, available at www.oceannet.org/medag/UKclimate-status/ClimateReport.html

Intergovernmental Oceanographic Commission (1993) *The Case for GOOS*, *IOC/INF-915 Corr*, Paris: UNESCO

Kildow, JT, Kite-Powell, H, Colgan, CS and Bruce, EJ (2000) 'Estimating the economic value of the ocean', *Sea Technology*, January 2000, 65–69

Kite-Powell, H and Colgan, C (2001) *The Economic Benefits of Coastal Ocean Observing Systems: The Gulf of Maine*, NOAA Office of Policy and Strategic Planning

Nicholls, N (2000) 'Opportunities to improve the use of seasonal climate forecasts', in Hammer, GL, Nicholls, N and Mitchell, C (eds) *Applications of Seasonal Climate Forecasting in Agricultural and Natural Ecosystems: The Australian Experience*, Dordrecht: Kluwer Academic Publishers

Portmann, JE (2000) *Review of Current UK Marine Observations in relation to Present and Future Needs*, Southampton: Inter-Agency Committee on Marine Science and Technology (IACMST)

Sassone, PG and Weiher, R (1999) 'Cost-benefit analysis of TOGA and the ENSO observing system', in Weiher, R, *Improving El Niño Forecasting: The Potential Economic Benefits*, NOAA Office of Policy and Strategic Planning

Science Applications International Corporation (2000) *Defining the Requirements of the US Energy Industry for Climate, Weather and Ocean Information*, NOAA Office of Oceanic and Atmospheric Research

Solow, AR, Adams, RM, Bryant, KJ, Legler, DM, O'Brien, JJ, McCarl, BA, Nayda WI and Weiher, R (1998) 'The value of improved ENSO prediction to US agriculture', *Climate Change* 39, 47–60

Texas A&M University (2002) *Scientific Design for the Common Module of the Global Ocean Observing System and the Global Climate Observing System: An Ocean Observing System for Climate*, available at http://ocean.tamu.edu

Weiher, RF (ed) (1999) *Improving El Niño Forecasting: The Potential Economic Benefits*, US Department of Commerce (NOAA Office of Policy and Strategic Planning)

Air quality monitoring

European Environment Agency (1999) 'Air pollution monitoring in Europe – problems and trends', *Topic Report No 26/1996*, available at http://reports.eea.eu.int/92-9167-058-8/en/tab_content_RLR

Fowler, D and Smith, R (2001) *Transboundary Air Pollution: Acidification, Eutrophication and Ground-Level Ozone in the UK*, available at www.nbu.ac.uk/negtap/finalreport.htm

Koelemeijer, R (2003) *Air & Climate Indicators for Supporting European Sustainable Development Policies*, available at www.gmes.info/library

Medina, S, Plasencia, A, Artazcoz, L, Quénel, P, Katsouyanni, K, Mücke, HG, De Saeger, E, Krzyzanowsky, M, Schwartz, J and the contributing members of the APHEIS group (2001) *APHEIS Air Pollution and Health: A European Information System. Final Scientific Report, 1999–2000*, Saint-Maurice: Institut de Veille Sanitaire

World Health Organization (2002) *The World Health Report 2002*, available at www.who.int/whr/2002/en

World Meteorological Organization/Global Atmosphere Watch (2001) *WMO/CEOS Report on a Strategy for Integrating Satellite and Ground-based Observations of Ozone*, available at http://netra1.wmo.ch/web/arep/reports/gaw140.pdf

Climate change research

Arnell, NW (1999) 'The effect of climate change on hydrological regimes in Europe: a continental perspective', *Global Environmental Change*, 9, 5–23

DEFRA (2001) *The UK's Third National Communication under the United Nations Framework Convention on Climate Change (UNFCCC)*, available at www.defra.gov.uk/environment/climatechange/3nc/pdf/climate_3nc.pdf

Estrela, T, Menéndez, M, Dimas, M, Marcuello, C, Rees, G, Cole, G, Weber, K, Grath, J, Leonard, J, Bering, N, Fehér, J, Lack, TJ and Thyssen, N (2001) 'Sustainable water use in Europe – part 3: extreme hydrological events: floods and droughts', *Environmental Issue Report No 21*, Luxembourg: Office of Official Publications of the European Union

European Commission (1999) *Towards a European Coastal Zone (ICZM) Strategy*, Luxembourg: Office of Official Publications of the European Union

Harrison, PA, Butterfield, RE and Downing, TE (1995) 'Climate change and agriculture in Europe: assessment of impacts and adaptations', *Research Report No 9*, Oxford: Environmental Change Unit, University of Oxford

Langford, IH and Bentham, G (1995) 'The potential effects of climate change on winter mortality in England and Wales', *International Journal of Biometeorology 39*, 141–47

Parry, ML (ed) (2000) *Assessment of Potential Effects and Adaptations for Climate Change in Europe: The Europe ACACIA Project*, Norwich: Jackson Environment Institute, University of East Anglia

United Nations Framework Convention on Climate Change (2003) 'The second report on the adequacy of the global observing systems for climate in support of the UNFCCC', *GCOS-82, WMO/TD No 1143*

US Climate Change Science Program and the Subcommittee on Global Change Research (2003) *Our Changing Planet – The Fiscal Year 2003 US Global Change Research Program and Climate Change Research Initiative*, available at www.usgcrp.gov/usgcrp/Library/ocp2003.pdf

REFERENCES

ACRES (2003) *Licence Conditions Covering ERS, JERS, RESURS, SPOT Data and Data Products Supplied by ACRES*, available at www.ga.gov.au/acres/referenc/enduser.pdf

Adams, RM, Bryant, KJ, McCarl, BA, Legler, DM, OBrien, JJ, Solow, A and Weiher, R (1995) 'Value of improved long-range weather information', *Contemporary Economic Policy* 13(3), 10–19

Agbu, PA and James, ME (1994) *The NOAA/NASA Pathfinder AVHRR Land Data Set Users Manual*, Greenbelt, MD: Goddard Distributed Active Archive Center

AGI (2004) *GI Gateway*, available at www.gigateway.org.uk

Ambite JL, Arens, Y, Hovy, E, Philpot, A, Gravano, L, Hatzivassiloglou, V, Klavans, J (2001) 'Simplifying data access: the energy data collection project', *Computer* 34(2), 47–54

ANZLIC (1999) *Policy Statement on Spatial Data Management: Towards the Australian Spatial Data Infrastructure*, Australia and New Zealand Land Information Council, Canberra, available at www.anzlic.org.au/policies.html

Arzberger P, Schroeder, P, Beaulieu, A, Bowker, G, Casey, K, Laaksonen, L, Moorman, D, Uhlir, P and Wouters, P (2004) 'An international framework to promote access to data', *Science* 303, 1777–78

BADC (2004) *British Atmospheric Data Centre*, available at http://badc.nerc.ac.uk

Barr, R (1998) 'The price of freedom', *GIS Europe* 7(3), 14–15

Beveridge, M, Howard, A, Burton, K and Holder, W (2003) 'The Ptolemy Project: a scalable model for delivering health information in Africa', *British Medical Journal* 327, 790–93

BIOPRESS (2004) *Linking Pan-European Land Cover Change to Pressures on Biodiversity*, available at www.creaf.uab.es/biopress

BMT (2001) *BNSC Market Sector Studies Programme – Marine Information Systems*, BMT Marine Information Systems Ltd, available at www.bnsc.gov.uk/index.cfm?fast=SSS_marine

Bugg, AL, Spencer, RD and Lee, A (2002) 'Applying GIS for developing regional forest agreements in Australia', *Photogrammetric Engineering and Remote Sensing* 68(3), 241–49

Cannizzaro, G (2003) *EUFOREO Main Service Reports Document*, available at www.cs.telespazio.it/earsc/EUFOREO

Carson, RT, Mitchell, RC, Hanemann, WM, Kopp, RJ and Ruud, PA (1992) *A Contingent Valuation Study of Lost Passive Use Values Resulting from the Exxon Valdez Oil Spill*, A Report to the Attorney General of the State of Alaska, US

Costanza, R, d'Arge, R, de Groot, R, Farber, S, Grasso, M, Hannon, B, Limburg, K, Naeem, S, O'Neill, RV, Paruelo, J, Raskin, RG, Sutton, P and van den Belt, M (1997) 'The value of the world's ecosystem services and natural capital', *Nature* 387, 253–60

DAAC (2004) *Distributed Active Archive Centre*, available at http://daac.gsfc.nasa.gov

DEFRA (2004) *Department for Environment, Food and Rural Affairs*, available at www.defra.gov.uk

Diepenbroek, M, Grobe, H, Reinke, M, Schindler, U, Schlitzer, R, Sieger, R and Wefer, G (2002) 'PANGAEA – an information system for environmental sciences', *Computers and Geosciences* 28, 1201–10

DISMAR (2004) *Data Integration System for Marine Pollution and Water Quality*, available at www.nersc.no/Projects/dismar

Dittert, N, Diepenbroek, M and Grobe, H (2001) 'Scientific data must be made available to all', *Nature* 414, 393

DLR (2004) *Deutsches Zentrum fur Luft-und Raumfahrt*, available at www.dlr.de/dlr

Dodge, M and Kitchin, R (2001) *Atlas of Cyberspace*, Harlow: Addison-Wesley

EC (2001a) *Communication from the Commission to the Council and the European Parliament: Global Monitoring for Environment and Security (GMES) Outline GMES EC Action Plan (Initial Period 2001–2003)*, Brussels, 23 October 2001, COM(2001) 609 final

EC (2001b) *Key Elements of the GMES EC Draft Action Plan: Initial Period 2001–2003*, European Commission, Brussels, 27 July 2001

EC (2002) *Infrastructure for Spatial Information in Europe*, available at http://inspire.jrc.it/home.html, document identifier INSPIRE DPLI PP v12-2 en

EC (2003a) *Consultation Paper on a Forthcoming EU Legal Initiative on Spatial Information for Community Policy-making and Implementation*, available at http://inspire.jrc.it/home.html

EC (2003b) *Report on the Feedback of the Internet Consultation on a Forthcoming EU Initiative Establishing a Framework for the Creation of an Infrastructure for Spatial Information in Europe*, available at http://inspire.jrc.it/reports/analysis_consultation_01092003.pdf

EC (2003c) *Draft Report on the Commission Communication on the Clean Air For Europe (CAFE) Programme: Towards a Thematic Strategy for Air Quality*, Brussels, COM(2001) 245 – C5-0598/2001 – 2001/2249 (COS)

EC (2003d) *Contribution to the Extended Impact Assessment of INSPIRE*, available at http://inspire.jrc.it/reports/fds_report_sept2003.pdf

EC (2004a) *Communication from the Commission to the European Parliament and the Council: Global Monitoring for Environment and Security (GMES): Establishing a GMES Capacity by 2008 – (Action Plan (2004–2008))*, Brussels, COM(2004) 65 final

EC (2004b) *INSPIRE Scoping Paper*, European Commission and European Environment Agency, document INSPIRE v1-02 en, 26 February 2004

EEA (1999a) *CORINE Catalogue of Data Sources – Annual Topic Update 1998*, available at http://reports.eea.eu.int/92-9167-144-4/en

EEA (1999b) *Information for Improving Europe's Environment*, Copenhagen, Denmark, available at http://org.eea.en.int/documents/brochure/brochurefull.pdf

EEA (2002) *Providing Policy Relevant Information for Europe's Environment – the EEA Strategy*, version 8.2, Copenhagen, Denmark

EEA (2003) *EEA Annual Work Programme 2003*, European Environment Agency, Copenhagen, Denmark

EEA (2004) *European Environment Agency*, available at www.eea.eu.int

Ehrlich, D, Estes, JE and Singh, A (1994) 'Applications of NOAA-AVHRR 1 km data for environmental monitoring', *International Journal of Remote Sensing* 15, 145–61

EMSC (2004) *European Mediterranean Seismological Centre*, available at www.emsc-csem.org

EO Summit (2003) *Earth Observation Summit*, available at www.earthobservationsummit.gov/index.html

ERS (2004) *The Economics of Food, Farming, Natural Resources, and Rural America*, Economic Research Service, available at www.ers.usda.gov

ESA (1998) *Envisat Data Policy*, ESA/PB-EO(97)57 rev 3, Paris, 19 February 1998

ESA (2003) *A New Perspective for Earth Observation: the Oxygen (O₂) Project*, Paris, available at http://esapub.esrin.esa.it/eoq/eoq71/suppl.pdf

ESA (2004a) *Earth Observation User Services*, available at http://earth.esa.int/services/catalogues.html

ESA (2004b) *ODISSEO catalogue*, available at http://odisseo.esrin.esa.it

EUMETSAT (2004) *The EUMETSAT Data Policy. User Guide to EUMETSAT Satellite Data*, available at www.eumetsat.de

EUROPA (2004) *Gateway to the European Union*, available at europa.eu.int/index_en.htm

European Council (2001) 'Council Joint Action of 20 July 2001 on the establishment of a European Union Satellite Centre', *Official Journal of the European Communities*, L200/5, 25 July 2001

EUROSTAT (2004) *Statistical Office of the European Union*, available at http://europa.eu.int/comm/eurostat

FAO (2004) *FAO Statistical Databases*, available at http://apps.fao.org/default.jsp

FGDC (2003) *Managing Historical Geospatial Data Records – Guide for Federal Agencies*, available at www.fgdc.gov/publications/documents/geninfo/histdata.pdf

FirstGov (2004) *Geodata.gov: US Maps and Data*, available at www.geodata.gov/gos

Flemming, NC (2001) 'EuroGOOS: analysis of the need for operational ocean remote sensing', *Operational Ocean Observations from Space*, EuroGOOS Publication 16, 6–12

G8 (2003) *Sommet d'Evian 2003*, summit reports available at www.g8.fr/evian/english/home.html

Garciá, EYG and Tuladhar, AM (2001) 'Development of a cadastral infrastructure in Guatemala: maximise data sharing, minimise data duplication', *GIM International* 15(8), 56–59

GeoConnections (2004) *Canadian Geospatial Data Infrastructure*, available at www.geoconnections.org/CGDI.cfm

Geography Network (2004) *Geography Network: Access a World of Information*, ESRI, available at www.geographynetwork.com

Global Internet Geography (2004) TeleGeography Research Group – PriMetrica, Inc, available at www.telegeography.com/products/index.php

GMFS (2004) *Global Monitoring for Food Security*, available at www.gmfs.info

GOOS (1998) *What is GOOS?*, available at http://ioc.unesco.org/goos/docs/whatis01.htm

Grindle, TJ and Burcham, FW (2002) 'Even minor volcanic ash encounters can cause major damage to aircraft', ICAO Journal 2, available at www.vnv-dalpa.com/opdebok/2002/20021103.html

GVM (2004) *Global Landcover 2000*, available at www.gvm.jrc.it/glc2000

Hall, M (2002) *Spatial Data Infrastructures in Australia, Canada and the United States*, Report elaborated in the context of a study commissioned by the EC in the framework of the INSPIRE initiative, available at http://inspire.jrc.it

Harris, R (1997) *Earth Observation Data Policy*, Chichester: John Wiley and Sons Ltd

Harris, R (2002) *Earth Observation Data Policy and Europe*, Lisse: AA Balkema

Hassall and Associates (1998) *Climate Change Scenarios and Managing the Scarce Water Resources of the Macquarie River: Final Report Prepared for the Australian Greenhouse Office*, Sydney: Hassall & Associates Pty Ltd

Houlding, SW (2001) 'XML – an opportunity for meaningful data standards in the geosciences', *Computers and Geosciences* 27, 839–49

IDC (2001) *A Proposal for a Commonwealth Policy on Spatial Data Access and Pricing*, Commonwealth Interdepartmental Committee on Spatial Data Access and Pricing, Canberra, Australia, available at www.osdm.gov.au/osdm/docs/Commonwealth_Policy_on_Spatial_Data_Access_and_Pricing.pdf

IFRC&RCS (2002) *World Disasters Report 2002 Focus on Reducing Risk*, Geneva, Switzerland, available at www.ifrc.org

IGFA (2002) *National Updates*, International Group of Funding Agencies for Global Change Research, presented at University of East Anglia, Norwich, 23–25 October 2002

INASP (2003) 'Optimising Internet bandwidth in developing country higher education', available at www.inasp.info/pubs/bandwidth

INFEO (2003) *INFEO Website*, available at www.eoportal.org, 1 July 2003

INSPIRE (2004) *Infrastructure for Spatial Information in Europe*, available at http://inspire.jrc.it/home.html

International Development Research Centre (2004) *Science for Humanity*, available at http://web.idrc.ca/en/ev-1-201-1-DO_TOPIC.html

Internet World Statistics (2004) *Internet Usage Statistics: The Big Picture*, available at www.internetworldstats.com/stats.htm

IOC (2001) *First Session of the Intergovernmental Working Group on IOC Oceanographic Data Exchange Policy*, document IOC/INF-1163, Paris, available at http://ioc3.unesco.org/iode/files/inf1163.pdf

IPCC (2001) *Third Assessment Report – Climate Change 2001*, available at www.ipcc.ch

Izzi Dien, M (2000) *The Environmental Dimensions of Islam*, Trowbridge: Redwood Books

Jaeger, PT (2003) 'The endless wire: e-government as global phenomenon', *Government Information Quarterly* 20, 323–31

Jennings, M (2002) *The African Internet: A Status Report*, available at www3.sn.apc.org/africa/afstat.htm

JRC (2004) *Digital Map Archive*, available at http://dma.jrc.it

Kite-Powell, H and Colgan, C (2001) *The Potential Economic Benefits of Coastal Ocean Observing Systems: The Gulf of Maine*, NOAA Office of Naval Research, available at www.publicaffairs.noaa.gov/worldsummit/pdfs/mainereport.pdf

Klinkenberg, B (2003) 'The true cost of spatial data in Canada', *The Canadian Geographer* 47(1), 37–49

Kondratyev, K and Cracknell, A (1998) *Observing Global Climate Change*, London: Taylor and Francis

Korte, G (1996) 'Weighing GIS benefits with financial analysis', *Government Finance Review*, October, 49–52

Landweber, L (1991) *International Internet Connectivity in 1991*, available at www.cybergeography.org/atlas/census.html

Landweber, L (1997) *International Internet Connectivity in 1997*, available at www.cybergeography.org/atlas/census.html

Macilwain, C (2000) 'Business lobby set to take EPA to court over data access', *Nature* 403, 236

MANHUMA (2000) *MANHUMA Website*, available at http://manhuma.planetek.it

Maps Direct (2004) *Maps Direct*, ESRI (UK), available at www.maps-direct.co.uk

Mattocks, S (2003) 'Information needs in relation to Petersberg tasks', 2nd GMES Forum, 14–16 January 2003, ESA-ESTEC, Noordwijk, The Netherlands

Mbarika, V, Jensen, M and Meso, P (2002) 'Cyberspace across sub-Saharan Africa', *Communications of the ACM* 45(12), 17–21

Muir, A and Oppenheim, C (2002) 'National information policy developments worldwide IV: copyright, freedom of information and data protection', *Journal of Information Science* 28(6), 267–481

Munich Re Group (2002) *Annual Review: Natural Catastrophes 2002*, available at www.munichre.com/pdf/topics_2002_e.pdf

Mutula, SM (2002) 'Internet connectivity and services in Kenya: current developments', *The Electronic Library* 20(6), 466–72

NOAA (2002) *Economic Statistics*, NOAA Office of Policy and Strategic Planning, available at http://205.156.54.206/pub/com/NOAAeconomicstatistics0402.pdf

Nowlin, WD, Briscoe, M, Smith, N, McPhaden, MJ, Roemmich, D, Chapman, P and Grassle, JF (2001) 'Evolution of a sustained ocean observing system', *Bulletin of the American Meteorological Society* 82(7), 1369–76

NSIDC (2004) *National Snow and Ice Data Center*, available at http://nsidc.org

OCEANIDES (2004) *Oceanides*, available at http://intelligence.jrc.cec.eu.int/marine/oceanides/oceanides.html

OECD (2004) *Science, Technology and Innovation for the 21st Century: Meeting of the OECD Committee for Scientific and Technological Policy at Ministerial Level, 29–30 January 2004 – Final Communique*, available at www.oecd.org

Ordnance Survey (1996) 'Economic aspects of the collection, dissemination and integration of government's geospatial information', a report arising from work carried out for Ordnance Survey by Coopers & Lybrand, Southampton: Ordnance Survey

OSCAR (2004) *Ocean Surface Current Analyses – Real time*, available at www.oscar.noaa.gov

Patin, S (1999) *Environmental Impact of the Offshore Oil and Gas Industry*, *Offshore Environment*, East Northport, NY: EcoMonitor Publishing

Pearce, D (1995) *Blueprint 4, Capturing Global Environmental Value*, London: Earthscan Publications Ltd

Peck, SC and Teisberg, TJ (1993) 'Global warming uncertainties and the value of information: an analysis using CETA', *Resource and Energy Economics* 15(1), 71–97

Perera, P (2003) 'A knowledge based european environment policy', 2nd GMES Forum, 14–15 January 2003, ESA-ESTEC, Noordwijk, The Netherlands

Planning Portal (2004) *Planning Portal*, available at www.planningportal.gov.uk

Remote Imaging Group (2004) 'MSG/Hot Bird dissemination overview', available at www.rig.org.uk/msgprgo4.htm

Roberts, AM and Moore, RV (1998) 'Data and databases for decision support', *Hydrological Processes* 12, 835–42

Robison, KK and Crenshaw, EM (2002) 'Post-industrial transformations and cyber-space: a cross-national analysis of Internet development,' *Social Science Research* 31, 334–63

ROSES (2004) *Real Time Ocean Services for Environment and Security*, available at http://roses.cls.fr/html/roses/services/links.html

Schreier, G (2002) 'Data policy implications on archive design – or vice versa?', in Harris, R (ed), *Earth Observation Data Policy and Europe*, Lisse: AA Balkema, 169–80

Scurlock, JMO, Asner, GP and Gower, ST (2001) *Global Leaf Area Index Data from Field Measurements, 1932–2000*, available at www.daac.ornl.gov, Oak Ridge National Laboratory Distributed Active Archive Center, Tennessee, US

Sears, G (2001) *Executive Summary: Geospatial Data Policy Study*, prepared for GeoConnections, Ottawa, Canada

Smith, PM, Kalluri, SNV, Prince, SD and DeFries, RS (1997) 'The NOAA/NASA Pathfinder AVHRR 8-km land data set', *Photogrammetric Engineering and Remote Sensing* 63, 27–32

Tang, W and Selwood, J (2003) *Connecting Our World: GIS Web Services*, California: ESRI Press

Tate, ED (1997) 'Access to information: the Canadian experience', *Journal of Information Science* 24(2), 75–82

UNECE (1996) *Health Costs Due to Road Traffic Related Air Pollution: An Impact Assessment Project of Austria, France and Switzerland*, available at www.unece.org

UNEP (2004) *GEO Data Portal*, available at http://geodata.grid.unep.ch

USGS (2003) *Architecture and Technology Program Offline Archive Media Trade Study*, available at http://edc.usgs.gov/archive/nslrsda

USGS (2004) *EarthExplorer*, available at http://edcsns17.cr.usgs.gov/EarthExplorer

Vandenbroucke, D and Peedell, S (2003) *GIS for Natura 2000: Example of the Need for a European Spatial Data Infrastructure*, available at http://wwwlmu.jrc.it/Workshops/8ec-gis/cd/papers/4_pa_dv.pdf

Vogt, PR, Carron, MJ, Jung, M-Y and Macnab, R (2003) *The Global Ocean Mapping Project (GOMAP) and UNCLOS: Optimizing Article 76 Surveys for Re-use in Portraying Global Bathymetry*, available at www.gmat.unsw.edu.au/ablos/ABLOS01Folder/VOGT.PDF

von der Dunk, F (2003) *The Galileo Project*, presentation to the European Centre for Space Law Workshop on New Trends in Space Law, University of Macerata, Italy

Wershler, T and Rancourt, J (2003) *The Dissemination of Government Geographic Data in Canada: Guide to Best Practice*, version 1.0, Ottawa, Canada: GeoConnections

WHO (2000) *Air Pollution Fact Sheet No 187*, Geneva: World Health Organization

WHO (2004a) *WHO Global Atlas of Infectious Diseases*, available at http://globalatlas.who.int

WHO (2004b) *WHO Statistical Information System*, available at http://www3.who.int/whosis/menu.cfm

Winker, K (1999) 'How to bring collections data into the net', *Nature* 401, 524

WMO (2003) *The Second Report on the Adequacy of the Global Observing Systems for Climate in Support of the UNFCCC*, GCOS-82, WMO/TD No 1143, available at www.wmo.ch/web/gcos/Second_Adequacy_Report.pdf

Woods, JD, Dahlin, H, Droppert, L, Glass, M, Vallerga, S and Flemming, NC (1996) *The Strategy for EuroGOOS*, EuroGOOS Publication No 1, Southampton: Southampton Oceanography Centre

Wright, TJ (2002) 'Remote monitoring of the earthquake cycle using satellite radar interferometry', available at www.earth.ox.ac.uk/~timw/papers/Wright_Visions_PTRS2002.pdf

WTO (2004) *World Trade Organization*, available at www.wto.org

WWF (2003) *EPO Activities: European Ecological Networks/Natura 2000*, available at www.panda.org

Index